Collins

Secure Maths
Year 5

a primary maths
intervention programme

Teacher's Pack

Collins

William Collins' dream of knowledge for all began with the publication of his first book in 1819.
A self-educated mill worker, he not only enriched millions of lives, but also founded a flourishing publishing house. Today, staying true to this spirit, Collins books are packed with inspiration, innovation and practical expertise. They place you at the centre of a world of possibility and give you exactly what you need to explore it.

Collins. Freedom to teach.
An imprint of HarperCollins*Publishers*
The News Building
1 London Bridge Street
London
SE1 9GF

Browse the complete Collins catalogue at
www.collins.co.uk

ISBN 978-0-00-822149-2

British Library Cataloguing in Publication Data
A catalogue record for this publication is available from the British Library.

Publishing manager Fiona McGlade
Editor Nina Smith
Project managed by Alissa McWhinnie, QBS Learning
Copyedited by Joan Miller
Proofread by Jo Kemp
Answers checked by Deborah Dobson
Cover design by Amparo Barrera and ink-tank and associates
Cover artwork by Amparo Barrera
Internal design by 2Hoots publishing services
Typesetting by QBS Learning
Illustrations by QBS Learning
Production by Rachel Weaver
Printed and bound by CPI

Contents

Introduction – Secure Year 5 Mathematics coverage

Domain	National Curriculum Attainment Target	Secure Maths Unit number	Page number
Number – number and place value	Read, write, order and compare numbers to at least 1 000 000 and determine the value of each digit	1	46
	Count forwards or backwards in steps of powers of 10 for any given number up to 1 000 000	2	48
	Interpret negative numbers in context, count forwards and backwards with positive and negative whole numbers, including through zero	3	50
	Round any number up to 1 000 000 to the nearest 10, 100, 1000, 10 000 and 100 000	4	52
	Read Roman numerals to 1000 (M) and recognise years written in Roman numerals	5	54
	Solve number problems and practical problems that involve all of the above	6	56
Number – addition and subtraction	Add and subtract whole numbers with more than four digits, including using formal written methods (columnar addition and subtraction)	7	58
	Add and subtract numbers mentally with increasingly large numbers	8	60
	Use rounding to check answers to calculations and determine, in the context of a problem, levels of accuracy	9	62
	Solve addition and subtraction multi-step problems in contexts, deciding which operations and methods to use and why	10	64

Domain	National Curriculum Attainment Target	Secure Maths Unit number	Page number
Number – multiplication and division	Identify multiples and factors, including finding all factor pairs of a number, and common factors of two numbers	11	66
	Know and use the vocabulary of prime numbers, prime factors and composite (non-prime) numbers	12	68
	Establish whether a number up to 100 is prime and recall prime numbers up to 19	13	70
	Multiply and divide numbers mentally, drawing upon known facts	14	72
	Multiply and divide whole numbers and those involving decimals by 10, 100 and 1000	15	74
	Multiply numbers up to four digits by a one- or two-digit number using a formal written method, including long multiplication for two-digit numbers	16	76
	Divide numbers up to four digits by a one-digit number using the formal written method of short division and interpret remainders appropriately for the context	17	78
	Recognise and use square numbers and cube numbers, and the notation for squared (^{2}) and cubed (^{3})	18	80
	Solve problems involving multiplication and division including using their knowledge of factors and multiples, squares and cubes	19	82
	Solve problems involving addition, subtraction, multiplication and division and a combination of these, including understanding the meaning of the equals sign	20	84
	Solve problems involving multiplication and division, including scaling by simple fractions and problems involving simple rates	21	86

Domain	National Curriculum Attainment Target	Secure Maths Unit number	Page number
Number – fractions (including decimals and percentages)	Identify, name and write equivalent fractions of a given fraction, represented visually, including tenths and hundredths	22	88
	Compare and order fractions whose denominators are all multiples of the same number	23	90
	Recognise mixed numbers and improper fractions and convert from one form to the other and write mathematical statements > 1 as a mixed number	24	92
	Add and subtract fractions with the same denominator and denominators that are multiples of the same number	25	94
	Multiply proper fractions and mixed numbers by whole numbers, supported by materials and diagrams	26	96
	Read and write decimal numbers as fractions	27	98
	Recognise and use thousandths and relate them to tenths, hundredths and decimal equivalents	28	100
	Round decimals with two decimal places to the nearest whole number and to one decimal place	29	102
	Read, write, order and compare numbers with up to three decimal places	30	104
	Solve problems involving number up to three decimal places	31	106
	Recognise the per cent symbol (%) and understand that per cent relates to 'number of parts per hundred', and write percentages as a fraction with denominator 100, and as a decimal fraction	32	110
	Solve problems which require knowing percentage and decimal equivalents of $\frac{1}{2}, \frac{1}{4}, \frac{1}{5}, \frac{2}{5}, \frac{4}{5}$ and those fractions with a denominator of a multiple of 10 or 25	33	112

Domain	National Curriculum Attainment Target	Secure Maths Unit number	Page number
Measurement	Convert between different units of metric measure (for example, kilometre and metre; centimetre and metre; centimetre and millimetre; gram and kilogram; litre and millilitre)	34	114
	Understand and use approximate equivalences between metric units and common imperial units such as inches, pounds and pints	35	116
	Measure and calculate the perimeter of composite rectilinear shapes in centimetres and metres	36	118
	Calculate and compare the area of rectangles (including squares), and including using standard units, square centimetres (cm^2) and square metres (m^2) and estimate the area of irregular shapes	37	120
	Estimate volume [for example, using 1 cm^3 blocks to build cuboids (including cubes)] and capacity [for example, using water]	38	122
	Solve problems involving converting between units of time	39	124
	Use all four operations to solve problems involving measure [money, length, mass, volume and money] using decimal notation, including scaling	40	126

Domain	National Curriculum Attainment Target	Secure Maths Unit number	Page number
Geometry – properties of shapes	Identify 3-D shapes, including cubes and other cuboids, from 2-D representations	41	130
	Know angles are measured in degrees: estimate and compare acute, obtuse and reflex angles	42	132
	Draw given angles, and measure them in degrees	43	134
	Identify: angles at a point and one whole turn (total 360°); angles at a point on a straight line and half a turn (total 180°); other multiples of 90°	44	136
	Use the properties of rectangles to deduce related facts and find missing lengths and angles	45	138
	Distinguish between regular and irregular polygons based on reasoning about equal sides and angles	46	140
Geometry – position and direction	Identify, describe and represent the position of a shape following a reflection or translation, using the appropriate language, and know that the shape has not changed	47	142
Statistics	Solve comparison, sum and difference problems using information presented in a line graph	48	146
	Complete, read and interpret information in tables, including timetables	49	148

Secure Maths is a structured intervention course for primary maths that can be followed in its entirety or dipped into as needed. It is a year-on-year programme, which can be used independently or alongside any programme or scheme.

The purpose of the series is to assist teachers in identifying children who are not on track to meet age-related expectations by the end of the school year, and to provide support to get them back on track, and ensure readiness for the next year or the SAT.

Secure Maths follows the 2014 programme of study and provides an Assess – Teach – Assess cycle.

For each year, there is a Teacher's pack and Pupil Resource Pack.

Teacher's pack

The Teacher's pack contains:

- 2 diagnostic tests with answers and gap analysis grids
- Units of teaching covering each National Curriculum Attainment Target (NCAT), containing background knowledge and teaching activities.

Diagnostic test 1

Diagnostic test 1

Number and place value

1. Continue this sequence:

   8800   8850   8900   ☐ ☐ ☐

2. Write 510 019 in words.

   _____

3. What is the value of the 3 in the number 436 729?

4. What year is MMXV?

5. Round 769 483 to the nearest 10 000.

6. These were the lowest temperatures recorded during one winter.

Place	Temperature (°C)
Anchorage	−9
London	1

   What was the temperature difference between Anchorage and London?

7. This mystery five-digit **odd** number rounds **down**. Read the clues. What number is it?

   It has the same number of tens as ten thousands.

   The thousands digit is three less than the ten thousands digit.

   Its hundreds digit is 6.

   The hundreds and tens digits add to 10.

Tests can be delivered by topic/domain or in their entirety.

Diagnostic tests highlight key areas of weakness and gaps in pupils' knowledge.

The questions in the test are at the end-of-year level expectation for the National Curriculum Attainment Target (NCAT) they are testing.

Answers to the diagnostic tests are provided immediately after the test, and are organised by topic/domain.

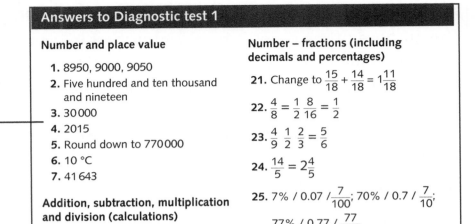

Answers to Diagnostic test 1

Number and place value

1. 8950, 9000, 9050
2. Five hundred and ten thousand and nineteen
3. 30000
4. 2015
5. Round down to 770000
6. 10 °C
7. 41 643

Addition, subtraction, multiplication and division (calculations)

8. a) 193 405
   b) 234 320
9. 100
10. 200 + 400 + 300 = 900

Number – fractions (including decimals and percentages)

21. Change to $\frac{15}{18} + \frac{14}{18} = 1\frac{11}{18}$

22. $\frac{4}{8} = \frac{1}{2}$ $\frac{8}{16} = \frac{1}{2}$

23. $\frac{4}{9}$ $\frac{1}{2}$ $\frac{2}{3} = \frac{5}{6}$

24. $\frac{14}{5} = 2\frac{4}{5}$

25. 7% / 0.07 / $\frac{7}{100}$; 70% / 0.7 / $\frac{7}{10}$; 77% / 0.77 / $\frac{77}{100}$

26. a) 4   b) 4.4

27. 7.489   7.49   7.5   7.511

28. 5.38

29. 7 out of 10 = 70%; 17 out of 20 =

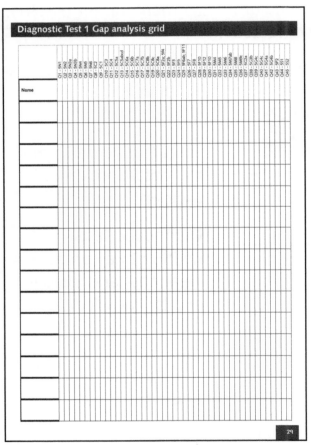

Gap analysis grids for both diagnostic tests are provided on the accompanying CD.

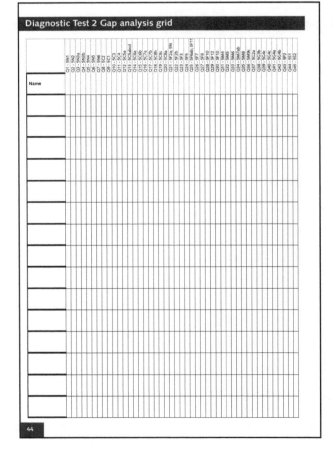

There is also a version in the Teacher's Guide on page 29 for Diagnostic test 1 and page 44 for Diagnostic test 2.

Each unit is linked to the National Curriculum domain and attainment target.

Prerequisites for learning list the knowledge children require before teaching of the unit can take place.

Learning outcomes

Background knowledge provides a clear and concise explanation of the mathematical concept(s) being covered in the unit.

Key vocabulary lists the mathematical words that children need to understand to access the unit.

Resources list: any physical resources required for teaching the unit.

Teaching activities are linked to objectives (above) and offer teaching plans for teachers/ TAs. Approximate timing is provided for each activity.

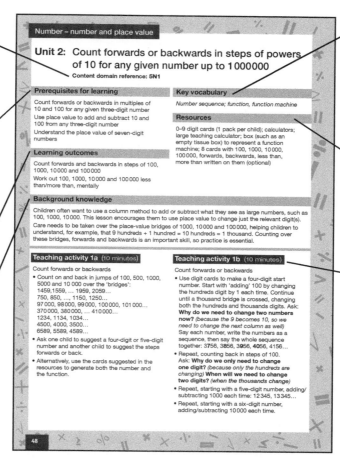

Number – number and place value

Unit 2: Count forwards or backwards in steps of powers of 10 for any given number up to 1 000 000
Content domain reference: 5N1

Prerequisites for learning

Count forwards or backwards in multiples of 10 and 100 for any given three-digit number
Use place value to add and subtract 10 and 100 from any three-digit number
Understand the place value of seven-digit numbers

Learning outcomes

Count forwards and backwards in steps of 100, 1000, 10 000 and 100 000
Work out 100, 1000, 10 000 and 100 000 less than/more than, mentally

Background knowledge

Children often want to use a column method to add or subtract what they see as large numbers, such as 100, 1000, 10 000. This lesson encourages them to use place value to change just the relevant digit(s).
Care needs to be taken over the place-value bridges of 1000, 10 000 and 100 000, helping children to understand, for example, that 9 hundreds + 1 hundred = 10 hundreds = 1 thousand. Counting over these bridges, forwards and backwards is an important skill, so practice is essential.

Key vocabulary

Number sequence; function, function machine

Resources

0–9 digit cards (1 pack per child); calculators; large teaching calculator; box (such as an empty tissue box) to represent a function machine; 8 cards with 100, 1000, 10 000, 100 000, forwards, backwards, less than, more than written on them (optional)

Teaching activity 1a (10 minutes)

Count forwards or backwards
• Count on and back in jumps of 100, 500, 1000, 5000 and 10 000 over the 'bridges':
1459,1559, … 1959, 2059…
750, 850, …, 1150, 1250…
97 000, 98 000, 99 000, 100 000, 101 000…
370 000, 380 000, … 410 000…
1234, 1134, 1034…
4500, 4000, 3500…
6589, 5589, 4589…
• Ask one child to suggest a four-digit or five-digit number and another child to suggest the steps forwards or back.
• Alternatively, use the cards suggested in the resources to generate both the number and the function.

Teaching activity 1b (10 minutes)

Count forwards or backwards
• Use digit cards to make a four-digit start number. Start with 'adding' 100 by changing the hundreds digit by 1 each time. Continue until a thousand bridge is crossed, changing both the hundreds and thousands digits. Ask: **Why do we need to change two numbers now?** *(because the 9 becomes 10, so we need to change the next column as well)* Say each number, write the numbers as a sequence, then say the whole sequence together: 3756, 3856, 3956, 4056, 4156…
• Repeat, counting back in steps of 100. Ask: **Why do we only need to change one digit?** *(because only the hundreds are changing)* **When will we need to change two digits?** *(when the thousands change)*
• Repeat, starting with a five-digit number, adding/ subtracting 1000 each time: 12 345, 13 345…
• Repeat, starting with a six-digit number, adding/subtracting 10 000 each time.

48

Resource sheets are often included to support teaching activities. These are photocopiable pages and can be found from page 150–182 of the Teacher's Guide.

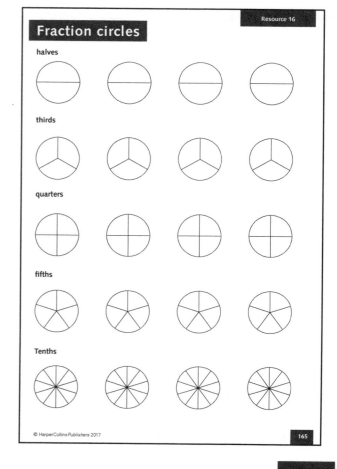

Fraction circles

halves

thirds

quarters

fifths

Tenths

© HarperCollins Publishers 2017

165

Answers to the practice and tests in the accompanying Pupil Resource Pack can be found be found from page 183–200 of the Teacher's Guide.

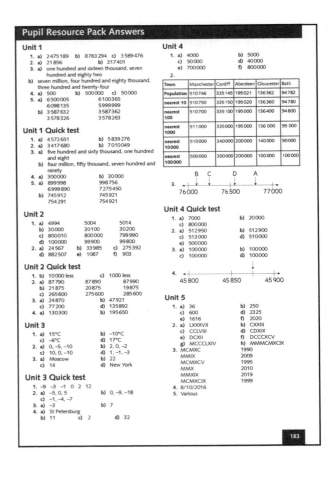

Pupil Resource Pack

The Pupil Resource Pack contains:

- Units, 1 per National Curriculum Attainment Target (NCAT), each containing:
  - Independent practice
  - Quick test.

Targeted practice linked to the teaching activities in the unit help children reinforce their learning.

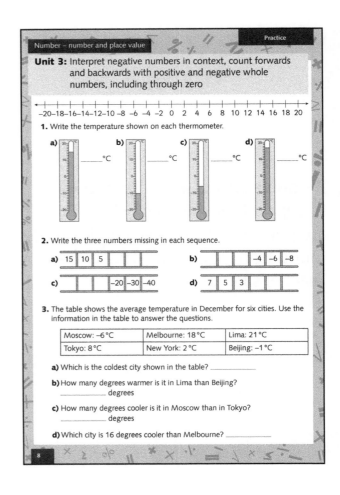

Quick tests enable teachers to check that children have mastered the objective and are ready to move on with the rest of the class. These can also be used for evidence.

How to use the programme

Assess

Diagnostic tests have been provided to assist you in identifying children in need of intervention, and to ascertain exactly what gaps there are in their knowledge. Tests can be set in their entirety, to get a picture of the children's knowledge of the year's programme of study, or they can be set by topic, to determine how children are faring in a particular domain of the NCAT.

Gap analysis

Use the gap analysis spreadsheet on the accompanying CD to quickly see where the gaps in class, and individual, knowledge are. Each question is testing a content domain, which are linked to units of teaching. By placing a Y against questions that children answered correctly, and an N against questions children answered incorrectly, the spread sheet will colour code each content domain either green or red. Red indicates a gap in knowledge which may require intervention.

A note: out of necessity, there are some instances where units may cover more than the child's requirements, for example, addition and subtraction where intervention is only required in subtraction. Please be mindful of this, and where possible, use the diagnostic test in conjunction with other data when identifying intervention requirements. As always, teacher judgement is by far the most powerful diagnostic tool.

Teach

Within each unit of teaching, we have identified specific objectives that the child needs to master in order to meet the requirements of that NCAT. For each objective, there is a choice of activity for the teacher/TA to use with the child or group of children. These provide alternative ways of teaching concepts, and cater for different learning styles. It is important that the intervention activity is different to the main teaching activity, and presents the concept in a different way to the child.

Some units cover just one learning outcome, but most will cover two outcomes and some even cover three outcomes. As each unit covers what would probably be taught over a week in class, some units could take more than one intervention session to cover. For each learning outcome there are two activities provided. The alternatives are slightly different ways of teaching each concept and can be used in an either/or approach or for further consolidation if children are still insecure after doing one of the activities. Using a limited variety of readily available resources, such as digit cards, dice, number lines and place value grids, helps children to visualise concepts without the distraction of a new resource to get used to. Activity (a), in many units, is the more efficient, concise method of teaching the outcome and can be used for children who are nearly there, but are making errors. Activity (b) tends to be an alternative or supplementary method that often uses different visual or concrete material support and comes at the learning outcome from a different angle. Activity (b) may often be a better starting point for those children who learn visually and cannot remember the 'rules' or for whom these rules make no sense or for those who have a number of gaps in their mathematics education. Some children could benefit from the related units in a previous year. Using the diagnostic tests of the previous year will help to identify areas where children need to catch up with basic concepts before they can successfully access the curriculum for their own year. Using the diagnostic tests of the previous year with a whole class at the beginning of a year could be useful tool in alerting the teacher to general gaps in understanding to inform future planning.

Practice

As with all learning, it is important for children to practice in order to reinforce the knowledge and skills acquired. Targeted practice is provided in the Pupil Resource Pack for each unit, linked directly to the intervention.

Assess

Quick tests have been provided for each unit, to be completed once the teacher is satisfied that the child has successfully completed the unit of intervention. This is a quick way of confirming that they have mastered the concept in question, and are ready to move on with the rest of the class. Quick tests can also be used as evidence of learning and achievement.

Diagnostic test 1

Number and place value

1. Continue this sequence:

8800   8850   8900   ☐   ☐   ☐

2. Write 510 019 in words.

_____

3. What is the value of the 3 in the number 436 729?

4. What year is MMXV?

5. Round 769 483 to the nearest 10 000.

6. These were the lowest temperatures recorded during one winter.

Place	Temperature (°C)
Anchorage	−9
London	1

What was the temperature difference between Anchorage and London?

7. This mystery five-digit **odd** number rounds **down**. Read the clues. What number is it?

It has the same number of tens as ten thousands.

The thousands digit is three less than the ten thousands digit.

Its hundreds digit is 6.

The hundreds and tens digits add to 10.

The ones digit is two more than the thousands digit.

Addition, subtraction, multiplication and division (calculations)

8. a) 248 092 − 54 687   **b)** 172 486 + 7009 + 54 825

9. 7476 + ☐ = 7576

10. Use rounding to 100 to estimate the answer to this calculation.

178 + 419 + 299 = ☐ + ☐ + ☐ = ☐

11. Each number is the sum of the two numbers below it. Complete the pyramid to find the value of *a*.

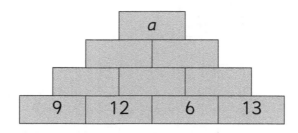

12. Fill in the factor spider diagram to identify the common factors of 20 and 30.

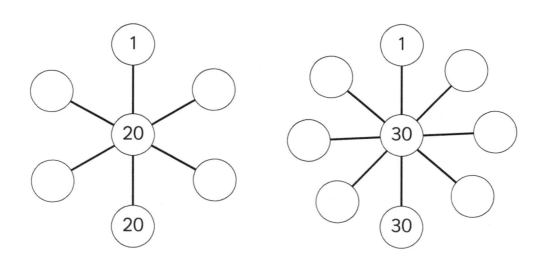

Common factors of 20 and 30 are: _____ _____ _____ _____ _____

13. Match the correct number from the box to the statement. **Use each number once**.

| 3 | 8 | 15 | 16 | 19 |

square number   _____

prime factor of 12   _____

cube number   _____

two-digit prime number   _____

multiple of 5   _____

14. $70 \times \boxed{} = 210$

15. $250 \div \boxed{} = 2.5$

16. Calculate: 2146×15.

17. Calculate:

$4\overline{)4756}$

18. Fill in the missing digit.

$(7 \times 4) + 2 = \boxed{} \times 6$

19. What is $\frac{2}{5}$ of 350?

20. A square has sides of 10 cm.
What is its area, in square centimetres?

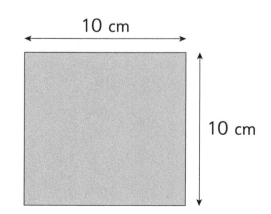

10 cm

10 cm

Number – fractions (including decimals and percentages)

21. Give your answer as a mixed number.

$$\frac{5}{6} + \frac{7}{9} = \underline{\hspace{8cm}}$$

22. Write the correct numbers to make these equivalents true.

$$\frac{\boxed{}}{8} = \frac{1}{2} \qquad \frac{8}{\boxed{}} = \frac{1}{2}$$

23. Write these fractions in order, smallest to largest.

$$\frac{2}{3} \qquad \frac{4}{9} \qquad \frac{5}{6} \qquad \frac{1}{2}$$

24. Give your answer as an improper fraction and as a mixed number.

$$7 \times \frac{2}{5} = \underline{\hspace{5cm}} \qquad \underline{\hspace{5cm}}$$

25. Write the sets of equivalent fractions, decimals and percentages.

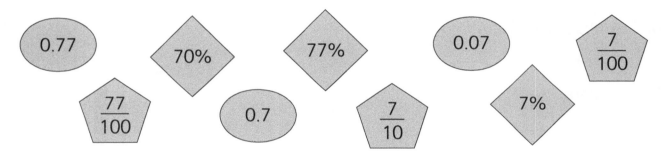

Fraction			
Decimal			
Percentage			

26. Round 4.38 to

   a) the nearest whole number _____

   b) the nearest tenth _____

27. Order the numbers, smallest to largest.

   7.5   7.489   7.511   7.49

_____

28. Two decimal numbers add together to equal 10. One of the numbers is 4.62.

   What is the other number?

29. Convert these scores to percentages.

   7 out of 10 = _____ = _____ %

   17 out of 20 = _____ = _____ %

   4 out of 5 = _____ = _____ %

   19 out of 25 = _____ = _____ %

   37 out of 50 = _____ = _____ %

30. Here is Lisa's shopping list. She pays with two £20 notes. How much change does she get?

S.No	Items	Prices
1.	Top	£11.75
2.	Skirt	£14.99
3.	Belt	£4.60

_____

Measurement

31. Show twenty-five to four on each clock face.

_____ a.m.

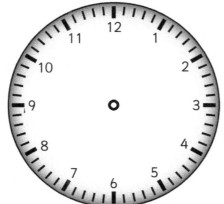

_____ p.m.

32. Choose numbers from the box to make these sentences true.

| 1 | 10 | 100 | 1000 |

a) ☐ cm = ☐ m

b) ☐ ml = ☐ litres

c) ☐ mm = ☐ cm

d) ☐ km = ☐ m

33. Complete the table to convert between pounds and kilograms.

kg	pounds
1	2.2
	4.4
10	
	44

34. Work out the area and the perimeter of this rectangular field.

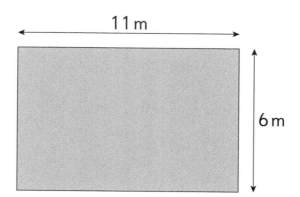

11 m

6 m

Area = _____ m²   Perimeter = _____ m

35. What is the volume of this cuboid?

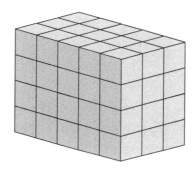

_____

36. Write these fractions of amounts in the unit of measure shown.

a) $\frac{1}{4}$ litre = _____ ml   **b)** $\frac{3}{4}$ m = _____ cm

c) $1\frac{9}{10}$ km = _____ m   **d)** $2\frac{1}{2}$ kg = _____ g

Geometry – properties of shapes

37. Tick (✓) and name the shapes that are regular polygons.

_____ _____ _____ _____ _____ _____

38. Name the shapes from their nets. Choose the correct name from the box.

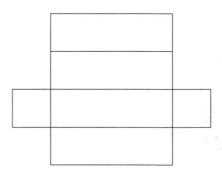 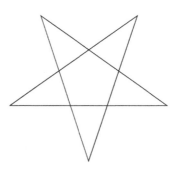

_____ _____

| cube | square-based pyramid | pentagonal-based pyramid |

| pentagonal prism | triangular prism | cuboid |

39. Measure the two angles shown. What is their sum?

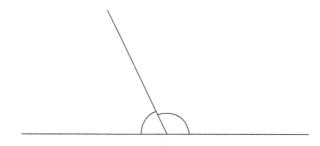

A = ⬚ °   B = ⬚ °   A + B = ⬚ °

40. Draw an angle of 155° at A.

A _____

22

41. Write the correct name for each angle.

| acute | right | reflex | obtuse |

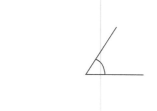   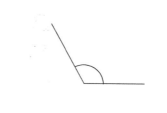

_____   _____

_____   _____

42. Work out the value of _a_.

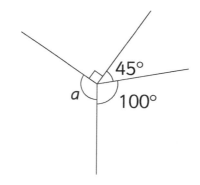

45°

_a_   100°

_a_ = _____ °

Geometry – position and direction

43. a) Reflect shape A in the line of reflection shown. Label it B.

   b) Translate (move) shape A two squares to the right and five squares down.
Label it C.

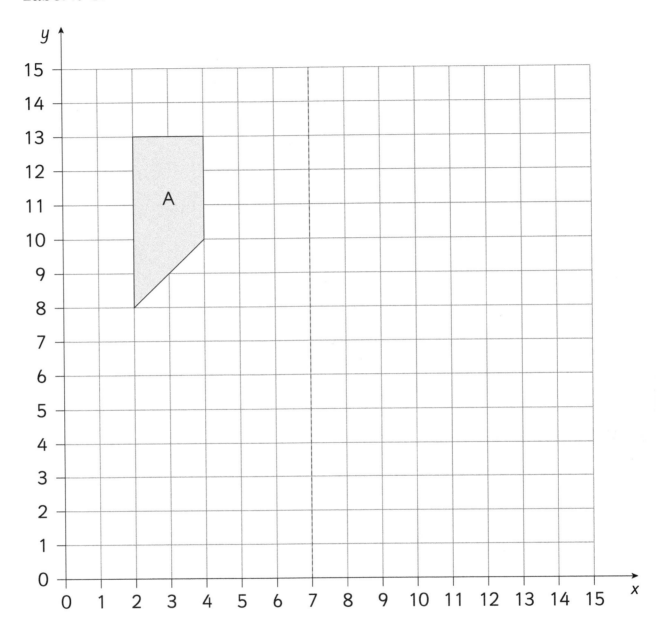

Statistics

44. Here is a timetable for a school swimming pool.

	Opening times	
	a.m.	p.m.
Monday	9:30–11:30	13:15–17:00
Tuesday	10:30–12:45	closed
Wednesday	closed	closed
Thursday	8:30–11:15	closed
Friday	9:30–12:00	13:30–15:45
Saturday	8:00–12:30	closed
Sunday	9:30–11:15	13:15–16:30

a) How long, in hours and minutes, is the pool open on Fridays?

_____

b) Max arrives at 11:15 on Tuesday morning.

How long can he swim for? _____

c) On which day is the pool open for the longest time? _____

45. The graph shows the average monthly temperature for Oslo, Norway.

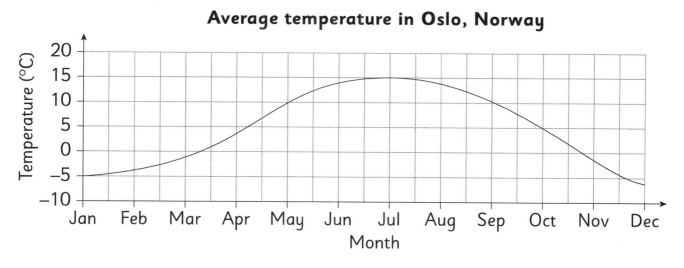

Average temperature in Oslo, Norway

a) What is the average May temperature in Oslo? _____ °

b) How many degrees colder is it in January than in July?

_____ degrees

Answers to Diagnostic test 1

Number and place value

1. 8950, 9000, 9050
2. Five hundred and ten thousand and nineteen
3. 30 000
4. 2015
5. Round down to 770 000
6. 10 °C
7. 41 643

Addition, subtraction, multiplication and division (calculations)

8. a) 193 405
   b) 234 320
9. 100
10. 200 + 400 + 300 = 900
11. 21, 18, 19   39, 37 $a = 76$
12. 20: 1 and 20, 2 and 10, 4 and 5
    30: 1 and 30, 2 and 15, 3 and 10, 5 and 6;
    common factors: 1, 2, 5, 10
13. Square number is 16; prime factor of 12 is 3; cube number is 8; prime number is 19; multiple of 5 is 15
14. 3
15. 100
16. 32 190
17. 1189
18. 5
19. 140
20. 100 cm^2

Number – fractions (including decimals and percentages)

21. Change to $\frac{15}{18} + \frac{14}{18} = 1\frac{11}{18}$
22. $\frac{4}{8} = \frac{1}{2}$ $\frac{8}{16} = \frac{1}{2}$
23. $\frac{4}{9}$ $\frac{1}{2}$ $\frac{2}{3} = \frac{5}{6}$
24. $\frac{14}{5} = 2\frac{4}{5}$
25. 7% / 0.07 / $\frac{7}{100}$; 70% / 0.7 / $\frac{7}{10}$; 77% / 0.77 / $\frac{77}{100}$
26. a) 4   b) 4.4
27. 7.489 7.49 7.5 7.511
28. 5.38
29. 7 out of 10 = 70%; 17 out of 20 = 85%; 4 out of 5 = 80%; 19 out of 25 = 76%; 37 out of 50 = 74%
30. £8.66

Measurement

31. 3:35 a.m.   15:35 p.m.

32. a) 100 cm = 1 m (or 1000 cm = 10 m)

b) 1000 ml = 1 litre

c) 10 mm = 1 cm (or 100 mm = 10 cm or 1000 mm = 100 cm)

d) 1 km = 1000 m

33.

kg	pounds
1	2.2
2	4.4
10	22
20	44

34. Area = 66 m², perimeter = 34 m

35. 60 cm³

36. a) $\frac{1}{4}$ litre = 250 ml

b) $\frac{3}{4}$ m = 75 cm

c) $1\frac{9}{10}$ km = 1900 m

d) $2\frac{1}{2}$ kg = 2500 g

Geometry – properties of shapes

37.

square   hexagon   equilateral triangle

38.

cuboid

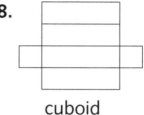

pentagonal-based pyramid

39. A = 65°   B = 115°   A + B = 180°

40. Accurate angle of 155°

41.

obtuse   right

acute   obtuse

42. $a = 125°$

Geometry – position and direction

43.

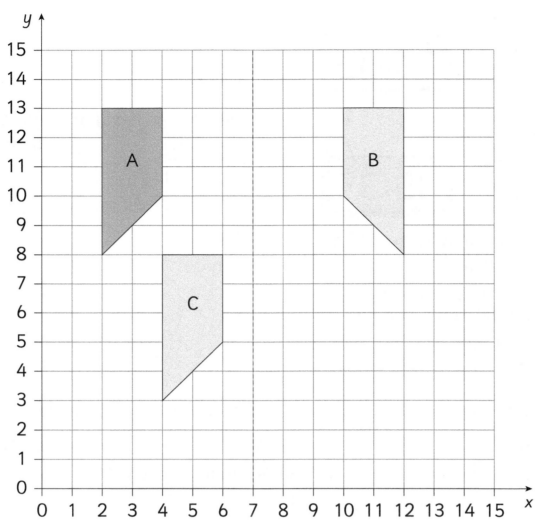

Statistics

44. a) 4 hours 45 minutes

    b) 1 hour 30 minutes

    c) Monday, 5 hours 45 minutes

45. a) 10°   **b)** 20 degrees

Name	Q1 - 5N1	Q2 - 5N2	Q3 - 5N3a	Q4 - 5N3b	Q5 - 5N4	Q6 - 5N5	Q7 - 5N6	Q8 - 5C2	Q9 - 5C1	Q10 - 5C3	Q11 - 5C4	Q12 - 5C5a	Q13 - 5C5abcd	Q14 - 5C6a	Q15 - 5C6b	Q16 - 5C7a	Q17 - 5C7b	Q18 - 5C8b	Q19 - 5C8c	Q20 - 5C8a	Q21 - 5F2a; 5f4	Q22 - 5F2b	Q23 - 5F3	Q24 - 5F5	Q25 - 5F6ab; 5F11	Q26 - 5F7	Q27 - 5F8	Q28 - 5F10	Q29 - 5F12	Q30 - 5F10	Q31 - 5M4	Q32 - 5M5	Q33 - 5M6	Q34 - 5M7ab	Q35 - 5M8	Q36 - 5M9c	Q37 - 5G2a	Q38 - 5G3b	Q39 - 5G4c	Q40 - 5G4c	Q41 - 5G4a	Q42 - 5G4b	Q43 - 5P2	Q44 - 5S1	Q45 - 5S2

Diagnostic test 2

Number and place value

1. Continue this sequence:

10 300   10 200   10 100

2. Write four hundred and sixty thousand and eleven in numerals.

_____

3. What is the number that has five tens, six thousands, four hundreds and three?

_____

4. Write CCXII in ordinary numbers.

_____

5. Round 218 945 to the nearest 1000. _____

6. Fill in the missing numbers on this number line.

 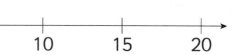

7. This table shows the average January temperatures for some cities.

Order the cities, coldest to warmest according to their January temperatures.

City	Average January temperature (°C)
Toronto	−6
Cairo	14
Moscow	−10
Paris	2
Beijing	0

Addition, subtraction, multiplication and division (calculations)

8. a) $819\,206 - 24\,863$ **b)** $78\,209 + 1753 + 473\,286$

9. What number is missing from the box?

$51\,980 - \boxed{} = 50\,980$

10. Use rounding to the nearest 100 to check if the answer to this calculation is correct.

$389 - 210 + 595 = 974$

11. Fill in the missing digits.

```
      ┌───┐     ┌───┐
   5  │   │  6  │   │
      └───┘     └───┘
  ┌───┐     ┌───┐
+ │   │  9  │   │  8
  └───┘     └───┘
─────────────────────
   9    6    8    2
─────────────────────
```

12. a) Write the factor pairs for 24 and 32.

24: 1, 24 _____ _____ _____

32: _____ _____ _____

b) Write down the common factors of 24 and 32.

_____ _____ _____ _____

13. Choose the correct number from the box to make each statement true.

Use each number once.

2	23	24	25	27

a) _____ is a square number.

b) _____ is a prime factor of 50.

c) _____ is a cube number.

d) _____ is a two-digit prime number.

e) _____ is a multiple of 6.

14. $1200 \div \boxed{} = 40$

15. Fill in the correct number to make this correct.

$\boxed{} \div 100 = 2.487$

16. Calculate 2903×21.

17. Calculate:

$5\overline{)7365}$

18. Fill in the missing digit.

$(35 \div 5) + 13 = \boxed{} \times 5$

19. a) How much does one DVD cost?

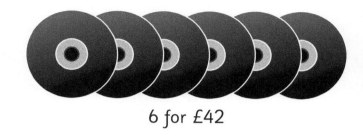

6 for £42

b) How much do four DVDs cost?

20. There are 28 children in class 5. List all the different ways the class can be split into equally sized groups.

Number – fractions (including decimals and percentages)

21. Give your answer as a mixed number.

$$\frac{2}{3} + \frac{8}{9} = \underline{\hspace{6cm}}$$

22. Draw lines to connect the equivalent fractions.

$$\frac{1}{4} \qquad \frac{3}{5} \qquad \frac{6}{12} \qquad \frac{4}{16} \qquad \frac{5}{10} \qquad \frac{6}{10}$$

23. Write these fractions in order, smallest to largest.

$$\frac{3}{4} \qquad \frac{1}{3} \qquad \frac{5}{8} \qquad \frac{1}{4}$$

_____

24. Give your answer as an improper fraction **and** as a mixed number.

$$3 \times \frac{5}{6} = \underline{\hspace{5cm}} = \underline{\hspace{5cm}}$$

25. Complete the table of equivalent fractions, decimals and percentages.

Fraction	Decimal	Percentage
$\frac{1}{4}$		
	0.9	
		20%
$\frac{1}{10}$		

26. Round 5.72 to **a)** the nearest whole number _____

   b) the nearest tenth.   _____

27. Write these athletes in order, highest jump to lowest jump.

Athlete	High-jump height (metres)
Arthur	1.299
Ben	1.39
Chris	1.349
Dan	1.4

First: _____

Second: _____

Third: _____

Fourth: _____

28. Over the weekend Lisa drank 4.5 litres of water.

David drank 3.725 litres.

How much more water did Lisa drink than David?

_____

29. Order these scores, best to worst, by converting them all to percentages.

72 out of 100   8 out of 10   15 out of 20   22 out of 25   39 out of 50

30. Finn buys one of each size pizza. What change does he get from £25?

Large pizza   £10.99

Medium pizza   £7.99

Small pizza   £4.75   _____

Measurement

31. A film starts at 17:50 and lasts $2\frac{1}{4}$ hours.
What time does it finish?

_____

32. Convert these units of measure.

    a) 300 cm = ◻ m   **b)** 4000 ml = ◻ litres

    c) 45 mm = ◻ cm   **d)** 7500 m = ◻ km

33. 1 litre = 1.75 pints. How many pints make 6 litres?

You can use this table to help you.

Litres	Pints
1	1.75

34. Work out the area and the perimeter of this square

Area = _____ m²

Perimeter = _____ m

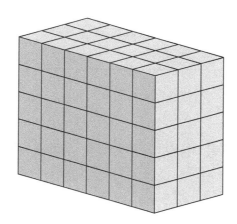

9 cm

35. What is the volume of this cuboid?

Volume = _____

36. Here are the ingredients to make 10 cookies.

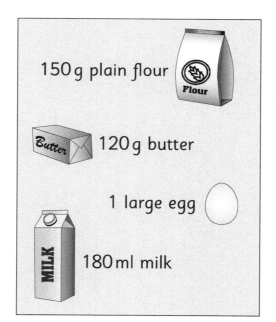

150 g plain flour

120 g butter

1 large egg

180 ml milk

a) How much flour is needed to make 20 cookies? _____

b) How much milk is needed to make 5 cookies? _____

c) How much butter is in each cookie? _____

Geometry – properties of shapes

37. a) Mark the right angles and the equal sides and angles of the **rectangle**.

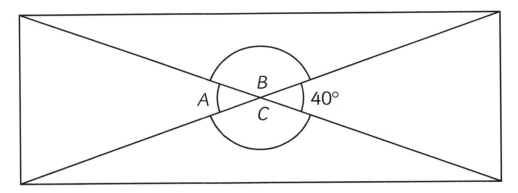

b) Write down the missing angles, where the diagonals cross each other.

A = _____°   B = _____°   C = _____°

38. Here are two nets for 3-D shapes. Write down the name of each shape.

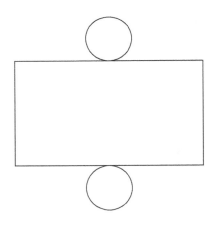

a) _____   **b)** _____

39. Measure angles *A* and *B* in the triangle. What is their sum?

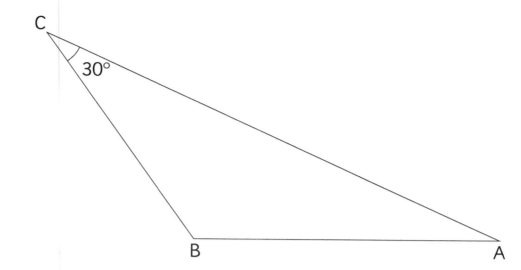

A = _____°   *B* = _____°

A + *B* + 30° = _____

40. Draw an angle of 45° at A.

_____ A

41. Choose the correct measurement for each angle from those given in the box.

The angles are not drawn to scale, so do not measure them.

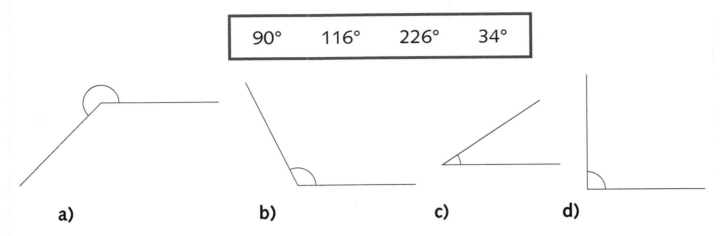

| 90° | 116° | 226° | 34° |

a)   b)   c)   d)

42. Work out the value of the angle marked *a*.

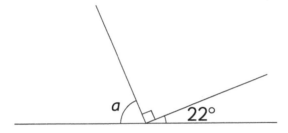

Geometry – position and direction

43. a) Reflect shape A in the line of reflection shown. Label it B.

b) Translate (move) shape A five squares to the right and six squares up
Label it C.

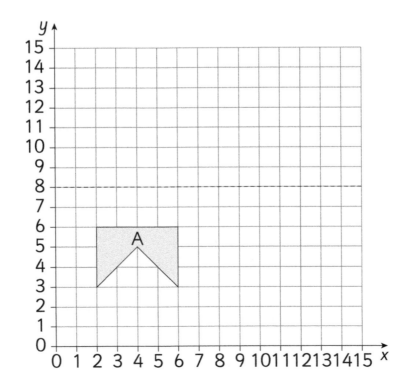

Statistics

44. This table shows the results of a survey on how children in Year 5 get to school.

	walk	bus	car	total
Boys	18			
Girls			13	28
Total	30	6	24	

a) Complete the table.

b) Which is the most popular way of getting to school for Year 5 boys?

_____

c) How many more boys than girls walk or go by car? _____

45. The graph shows the average heights of boys.

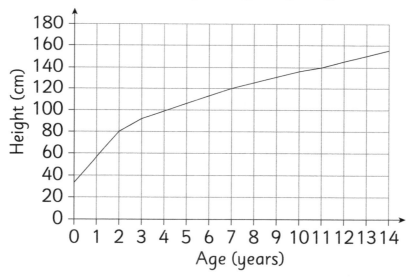

Average height of boys

a) What is the average height of a 10-year-old boy? _____

b) Billie is average height. He is 1 metre tall. How old is he?

_____ years old.

Number and place value

1. 10 000; 9900; 9800
2. 460 011
3. 6453
4. 212
5. 219 000
6. –15, –10, –5, 0, 5
7. Moscow, Toronto, Beijing, Paris, Cairo

Addition, subtraction, multiplication and division (calculations)

8. a) 794 343   b) 553 248
9. 1000
10. 400 – 200 + 600 = 800. The answer is incorrect.
11. 5764 + 3918 = 9682
12. a) 24 is 1, 24; 2, 12; 3, 8; 4, 6
    32 is 1, 32; 2, 16; 4, 8
    b) 1, 2, 4, 8
13. Square number is 25; prime factor of 50 is 2; cube number is 27; prime number is 19 or 23; multiple of 6 is 24
14. 30
15. 248.7
16. 60 963
17. 1473
18. 4
19. a) £7   b) £28
20. 2 of 14, 4 of 7, 7 of 4, 14 of 2

Number – fractions (including decimals and percentages)

21. $1\frac{5}{9}$

22.

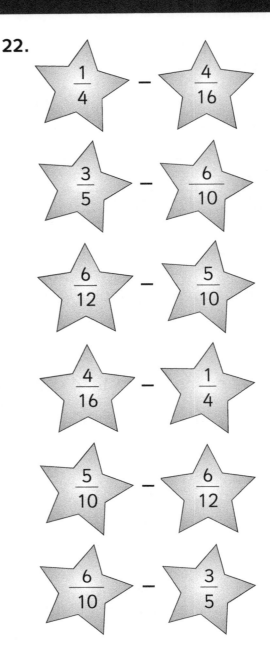

23. $\frac{1}{4}$ $\frac{1}{3}$ $\frac{5}{8}$ $\frac{3}{4}$

24. $\frac{15}{6} = 2\frac{1}{2}$

25.

Fraction	Decimal	Percentage
$\frac{1}{4}$	0.25	25%
$\frac{9}{10}$	0.9	90%
$\frac{1}{5}$	0.2	20%
$\frac{1}{10}$	0.1	10%

26. a) 6   **b)** 5.7

27. First: Dan, second: Ben, third: Chris, fourth: Arthur

28. 0.775 litres

29. 22 out of 25 = 88%; 8 out of 10 = 80%; 39 out of 50 = 78%; 15 out of 20 = 75%; 72 out of 100 = 72%

30. £1.27

Measurement

31. 20:05 or 8:05 p.m.

32 a) 300 cm = 3 m

   b) 4000 ml = 4 litre

   c) 45 mm = 4.5 cm

   d) 7500 m = 7.5 km

33. Table: for example, 2 l litres = 3.5, 3 l litres = 5.25, 4 l litres = 7, 5 l litres = 8.75, 6 l litres =10.5, 6 litres = 10.5 pints

34. Area= 81 m², Perimeter = 36 m

35. 90 cm³

36. a) 300g   **b)** 90ml   **c)** 12g

Geometry – properties of shapes

37. a)

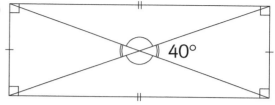

   b) 40°, 140°, 140°

38. Cylinder and triangular prism

39. A 25°, B 125°; A + B = 150°

40. Accurate angle of 45°

41. a 226°; b 116°; c 34°; d 90°

42. $a = 68°$

Geometry – position and direction

43.

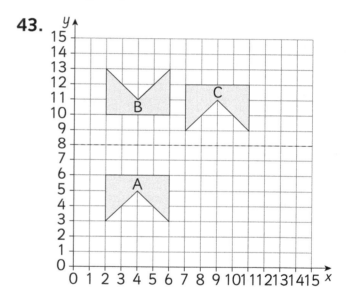

Statistics

44. a)

	walk	bus	car	total
Boys	18	3	11	32
Girls	12	3	13	28
Total	30	6	24	60

   b) walk   **c)** 4

45. a) 130–140 cm

   b) about 4

Diagnostic Test 2 Gap analysis grid

Name	Q1 - 5N1	Q2 - 5N2	Q3 - 5N3a	Q4 - 5N3b	Q5 - 5N4	Q6 - 5N5	Q7 - 5N6	Q8 - 5C2	Q9 - 5C1	Q10 - 5C3	Q11 - 5C4	Q12 - 5C5a	Q13 - 5C5abcd	Q14 - 5C6a	Q15 - 5C6b	Q16 - 5C7a	Q17 - 5C7b	Q18 - 5C8b	Q19 - 5C8c	Q20 - 5C8a	Q21 - 5F2a; 5f4	Q22 - 5F2b	Q23 - 5F3	Q24 - 5F5	Q25 - 5F6ab; 5F11	Q26 - 5F7	Q27 - 5F8	Q28 - 5F10	Q29 - 5F12	Q30 - 5F10	Q31 - 5M4	Q32 - 5M5	Q33 - 5M6	Q34 - 5M7ab	Q35 - 5M8	Q36 - 5M9c	Q37 - 5G2a	Q38 - 5G3b	Q39 - 5G4c	Q40 - 5G4c	Q41 - 5G4a	Q42 - 5G4b	Q43 - 5P2	Q44 - 5S1	Q45 - 5S2

Units

Unit 1: Read, write, order and compare numbers to at least 1 000 000 and determine the value of each digit

Content domain reference: **5N2** and **5N3a**

Prerequisites for learning

Read, write, order and compare three-digit numbers

Recognise the place value of each digit in a four-digit number

Learning outcomes

Read and write numbers to at least 1 000 000, knowing the value of each digit

Order numbers to at least 1 000 000, knowing the value of each digit

Key vocabulary

Million, hundred thousand, ten thousand

Resources

Resource 1: Place-value grids; 0–9 dice or number cards 0–9

Background knowledge

Our number system is a decimal system, which means that it is base ten. Each 'place' has 10 × the value of the place to the right of it. 10 000 = 10 × 1000; 1 000 000 = 10 × 100 000.

The value of each digit is determined by its position in the number.

The digits in larger numbers (with five or more digits) are grouped in threes, separated by spaces: 1 234 567. This means that, as long as children can read a three-digit number, they should be able to read any larger number.

Note: Sometimes children will see the groups separated by commas, for example, 1,234,567, but this format is being phased out in line with the International System of Units (*Système international d'unités*, SI), which is a metric system.

Confusion often arises where a number has several zeros as place holders, such as 2 000 015 and four hundred thousand and six. Help children to consider how many digits are needed: a number beginning with 'million' will be a seven-digit number, with the million first followed by a space, then three groups of three digits, also separated by spaces: _ _ _ _ _ _ _ , no more, no less. Using digit cards in a large place-value grid could also help, with children saying the number, putting the card in the correct column and filling any spaces with zeros.

In **whole** numbers, the larger the number of digits, the larger the number. A six-digit whole number will always be larger than any five-digit whole number. However, as this is not true with decimal places (4·5 is larger than 4·499 for example), do not labour this point.

Teaching activity 1a (15 minutes)

Read and write numbers to at least 1 000 000

- Start by counting with the children, in multiples of 1, 2, 5 or 10, on and back, over the hundreds and thousands place-value bridges, choosing different starting numbers, for example, 397, 985, 5006, 2040. Children could choose a start number, directed by you: *a five-digit number; a four-digit multiple of 10...*
- Ensure that children can read and write three-digit numbers, then explain that if they can

do this, they will be able to read much larger numbers.

- Write some four-digit numbers with a gap after the thousands digit and no zero place holders, explaining that the gap should be read as the word 'thousand'. Repeat with a few five-digit, then six-digit numbers with no zeros at first. Now include some numbers with zeros, such as 1006, 41 017, 730 025, 800 009. Ask: **How do we say this number?** *(eight hundred thousand and nine)* **Which word do we use for this space?** *(thousand)*

- Say some four- to six-digit numbers for children to write on mini-whiteboards, in numerals (explain that numerals are numbers written as 1, 2, 3, for example). Include some of the possible problem numbers, as before, checking carefully that they are not writing extra zeros or putting the gap in the wrong place. Remind them that there must be three digits after a gap. Ask: **Have you written three numbers after the gap?**

- Repeat with seven-digit numbers, explaining that they now need two spaces, so that the final six digits are in two groups of three. Say: **The first gap now represents the word 'million'.** Read and write some seven-digit numbers together, including some awkward ones with zeros in different places: 3 286 102; 6 080 945; 5 100 009; 2 006 090; 8 000 014. Remind children that there must be three digits between the spaces and after the final space. Ask: **What is the value of the [6] in this number?**

Teaching activity 1b (15 minutes)

Read and write numbers to at least 1 000 000

- Roll a 0–9 dice to generate numbers to form four- to seven-digit numbers, beginning with smaller numbers and increasing as appropriate. Write these digits in one of the Place-value grids (Resource 1: Place-value grids), deciding where the first digit needs to go (in the ten thousands column for a five-digit number, the hundred thousands column for a six-digit number etc.). This will help to reinforce, for example, that in a five-digit number the highest place value is tens of thousands. Use the grid headings to help children to read the number.

- Say some four- to seven-digit numbers in words and ask children to write the number in the grid, including some with zeros as place holders. Reinforce the 'groups of three' idea, stressing that there must be three digits after each space, so some places may need to be filled with zeros, for example, say: **If there are no hundreds, then you must write 0 under the hundreds heading.**

Teaching activity 2a (15 minutes)

Order numbers to at least 1 000 000, knowing the value of each digit

- Use the Place-value grids (Resource 1: Place-value grids) to record several numbers of different lengths. Ask individuals to give you any four-digit, five-digit, six-digit and seven-digit number. Then ask for specifics such as: a four-digit number with five hundreds; a six-digit even number; a five-digit multiple of 10; a six-digit number with no hundreds…

- Discuss the value of certain digits in some of the numbers. Ask: **What is the value of the 3 in 230 981?... 123 456? Which is the largest number? How do you know?**

- Write the numbers in order, smallest to largest or largest to smallest.

- Repeat with a set of numbers of the same length, with the same first and second digits. Ask: **Which number is the largest? How do you know? Which digit do you need to look at? Order them.**

Teaching activity 2b (15 minutes)

Order numbers to at least 1 000 000, knowing the value of each digit

- Generate sets of four, four-digit numbers, using dice or number cards, writing them on rectangles of paper or card so that they can be easily moved around. Write < and > on pieces of card and pick out pairs of numbers to compare, for example, 3456 < 3546.

- Rearrange the cards so that the numbers are in order, from smallest to largest. Repeat with a different set of numbers and order them from largest to smallest.

- Repeat with sets of four, five-digit, six-digit then seven-digit numbers to compare and order.

Unit 2: Count forwards or backwards in steps of powers of 10 for any given number up to 1000000

Content domain reference: 5N1

Prerequisites for learning

Count forwards or backwards in multiples of 10 and 100 for any given three-digit number

Use place value to add and subtract 10 and 100 from any three-digit number

Understand the place value of seven-digit numbers

Learning outcomes

Count forwards and backwards in steps of 100, 1000, 10000 and 100000

Work out 100, 1000, 10000 and 100000 less than/more than, mentally

Key vocabulary

Number sequence; function, function machine

Resources

0–9 digit cards (1 pack per child); calculators; large teaching calculator; box (such as an empty tissue box) to represent a function machine; 8 cards with 100, 1000, 10000, 100000, forwards, backwards, less than, more than written on them (optional)

Background knowledge

Children often want to use a column method to add or subtract what they see as large numbers, such as 100, 1000, 10000. This lesson encourages them to use place value to change just the relevant digit(s).

Care needs to be taken over the place-value bridges of 1000, 10000 and 100000, helping children to understand, for example, that 9 hundreds + 1 hundred = 10 hundreds = 1 thousand. Counting over these bridges, forwards and backwards is an important skill, so practice is essential.

Teaching activity 1a (10 minutes)

Count forwards or backwards

• Count on and back in jumps of 100, 500, 1000, 5000 and 10 000 over the 'bridges':
1459,1559, … 1959, 2059…
750, 850, …, 1150, 1250…
97 000, 98 000, 99 000, 100 000, 101 000…
370 000, 380 000, … 410 000…
1234, 1134, 1034…
4500, 4000, 3500…
6589, 5589, 4589…

• Ask one child to suggest a four-digit or five-digit number and another child to suggest the steps forwards or back.

• Alternatively, use the cards suggested in the resources to generate both the number and the function.

Teaching activity 1b (10 minutes)

Count forwards or backwards

• Use digit cards to make a four-digit start number. Start with 'adding' 100 by changing the hundreds digit by 1 each time. Continue until a thousand bridge is crossed, changing both the hundreds and thousands digits. Ask: **Why do we need to change two numbers now?** *(because the 9 becomes 10, so we need to change the next column as well)* Say each number, write the numbers as a sequence, then say the whole sequence together: 3**7**56, 3**8**56, 3**9**56, **40**56, 4156…

• Repeat, counting back in steps of 100. Ask: **Why do we only need to change one digit?** *(because only the hundreds are changing)* **When will we need to change two digits?** *(when the thousands change)*

• Repeat, starting with a five-digit number, adding/subtracting 1000 each time: 12 345, 13 345…

• Repeat, starting with a six-digit number, adding/subtracting 10 000 each time.

• Check understanding by counting on and back in steps of 100 in five- and six-digit numbers and of 1000 in six-digit numbers. Ask: **Can you say the next number before I change the digit card?**

Teaching activity 2a (15 minutes)

Work out 100, 1000, 10000 and 100 000 less than/more than

• Shuffle two packs of digit cards together and use them to generate five- and six-digit numbers. Explain that children are going to work out 100, 1000, 10000 and 100000 less than/more than this number by changing just one digit. Say: **Sometimes you will need to change two digits, though. Can you spot the pattern for when this happens?**

• Use another set of digit cards to change the relevant digit. Ask: **What number is 1000 less than this one? Which digit needs to change? What number is 10000 more than this one?** Make the link with addition and subtraction. Say: **Finding 10000 more than is the same as adding 10000; finding 1000 less than is the same as subtracting 1000.** Ask: **Why do you not need to write this down to add/subtract?**

• Check whether children can identify the numbers without using the digit cards. If they cannot, they should use the digit cards to complete the practice page.

Teaching activity 2b (15 minutes)

Work out 100, 1000, 10000 and 100000 less than/more than

• Use digit cards to generate four-to six-digit numbers, placing them to the left of the 'function machine box'. Place the large calculator on top of the box, explaining that this machine, which is called a function machine, is going to change the number in some way. Start with 100 more/less. Say: **The function machine is going to swallow this number and when it comes out, it will be 100 more.** Enter the number on the calculator and add 100. Show the output number. For example, 27435 goes into the machine and 27**5**35 comes out. Repeat with other input numbers.

• Repeat for 100 less, 1000 more/less, 10000 more/less.

• Now put an input number on the left of the box and an output number on the right to challenge children to work out what the calculator has done; for example, 45692 in and 44692 out means the calculator has subtracted 1000. Give the children practice with straightforward calculations, as necessary, before challenging them with some operations that go over place-value bridges.

Input	Output	Function
2996	3096	+ 100
23908	24008	+ 100 (not + 1000)
399800	400800	+ 1000
502375	492375	– 10000

• Finally, draw some function machines, like those below, asking children to work out the output number, to prepare them for the practice page.

38706 | + 1000 | ?   185099 | – 100 | ?

Unit 3: Interpret negative numbers in context, count forwards and backwards with positive and negative whole numbers, including through zero

Content domain reference: 5N5

Prerequisites for learning

Count backwards through zero
Know that zero is a number so must be counted

Key vocabulary

Negative, zero, temperature, degrees, Celsius, thermometer

Learning outcomes

Count backwards and forwards across zero
Use negative numbers in the context
of temperature

Resources

Resource 2: Negative number lines –20 to 20;
Resource 3: Thermometers; dice with more than six sides

Background knowledge

A negative number describes a number less than/below zero. Numbers above zero are positive numbers. Zero is neither positive nor negative, but it is a number and must be counted and not jumped over. A number line with matching ranges of positive and negative numbers is symmetrical around the zero: –3 is **more** than –4 as it is closer to zero. This is opposite to what happens with positive numbers, where 3 < 4. You read –3 as 'negative 3'. It is often called 'minus 3'.

Teaching activity 1a (10 minutes)

Count backwards and forwards across zero

- As a group, practise counting back in 1s from 10 to –10, then in multiples of 2, 3, 4, 5 and 10, starting with a multiple of the 'counting in' number so that zero is always in the count.

- Repeat, starting with a negative multiple: –25 for 5s, –60 for 10s.

- Now count back and forwards in 2s from an odd number: 5, 3, 1, –1, –3...; –11, –9, ... Ask: **Why is zero not in our count?** *(Because it is not in the sequence of counting in 2s)* **Why is zero not in the sequence?** *(Because the sequence is odd numbers and zero is being counted as an even number)*

- Repeat in steps of 5 and 10 from a 'non-multiple': 21, 16, 11, 6, 1, –4...

- Remind children that zero must always be counted in the count, even if it is not in the sequence.

Teaching activity 1b (10 minutes)

Count backwards and forwards across zero

- Children complete all the lines on Resource 2: Number lines by filling in the missing numbers.

- Use the completed number lines and dice to play this game.

  - Choose a positive start number less than 5.

  - Roll the dice to decide how many to count back.

  - Count back along the number line.

  - Write the move, for example, $4 - 6 = -2$, or simply record it on the number line, saying: **4 take away 6 is negative 2.**

  - Choose a negative start number greater than –5.

  - Roll the dice to decide how many to count forwards.

  - Write the move as, for example, $-4 + 6 = 2$, or record on the number line, saying: **–4 add 6 is positive 2.**

  - Repeat with a different positive starting number, but roll the dice **twice** and keep moving in the same (negative/left) direction, recording each move separately as a subtraction or by drawing on the number lines: $4 - 6 = -2$; $-2 - 3 = -5$.

- Repeat, starting with a number less than −10 and rolling the dice twice or more. Keep moving in the same (positive/right) direction: −15 + 4 = −11; −11 + 6 = −5; −5 + 7 = 2…
- Use the number lines to count on and back in steps of 2 from an even number, then from an odd number, then in steps of 5 from a multiple of 5, then a non-multiple of 5.

Teaching activity 2a (15 minutes)

Use negative numbers in the context of temperature

- Referring to Resource 3: Thermometers, explain that temperature is measured in degrees. The higher the number, the higher the temperature (the hotter it feels); the lower the number, the lower the temperature (the cooler it feels). 0 °C is the freezing point of water so temperatures below 0 °C feel very cold.
- Look carefully at the thermometers on Resource 3. Ask children to point to various temperatures: 20 °C, −15 °C, 45 °C, −35 °C.
- Ask children to fill in some temperatures on Resource 3 by colouring the thermometer gauge. They should choose two positive and two negative temperatures and write the temperature, in numerals, under each thermometer. Let them check each other's work. Ask: **Has your partner recorded the correct temperature? Which temperatures are below zero? Which is the highest/lowest temperature?**
- Ask children to write their temperatures in order: highest to lowest or lowest to highest.

Teaching activity 2b (15 minutes)

Use negative numbers in the context of temperature

- Use Resource 3: Thermometers to support answers to some questions related to change in temperature, to practise the language. Include moving from positive to negative, negative to positive and negative to negative. Ask each question several times with different values, starting with multiples of 5 and 10. Then move on to use any other numbers within the range of the resource (−50 to 50).
- Ask questions such as:
  - **The temperature has gone up from −10 °C to 5 °C. How many degrees higher is it?** (15)
  - **The temperature has gone down from 15 °C to −10 °C. How many degrees lower is it?** (25)
  - **One day the daytime temperature was 10 °C. Overnight it fell by 15 degrees. What was the night-time temperature?** (−5 °C)
  - **In London the temperature one day was 15 °C. In Moscow it was 10 degrees lower. What was the temperature in Moscow?** (5 °C)
  - **What temperature is 20 degrees lower than 15 °C?** (−5 °C)
  - **What temperature is 15 degrees higher than −5 °C?** (10 °C)
  - **What is the difference in temperature between 2 °C and −7 °C?** (9 degrees)
  - **What is the difference in temperature between −3 °C and −8 °C?** (5 degrees)
  - **In Alaska the average temperature in winter is −8 °C; in Spain it is 12 °C . How many degrees warmer is it in Spain?** (20)
  - **The temperature has risen from −5 °C to 5 °C. How many degrees is this?** (10)
  - **The temperature has fallen from −5 °C to −15 °C. How many degrees is this?** (10)
  - **In Bristol the temperature was 11 °C. In New York it was −2 °C. What was the temperature difference?** (13 °C)

Unit 4: Round any number up to 1 000 000 to the nearest 10, 100, 1000, 10 000 and 100 000

Content domain reference: 5N4

Prerequisites for learning

Understand the rules of rounding

Round two-digit numbers to the nearest 10, three-digit numbers to nearest 100

Learning outcomes

Round four-digit, five-digit and six-digit numbers to the nearest 1000, 10 000 and 100 000 respectively

Round any number to the nearest 10, 100, 1000, 10 000 and 100 000

Key vocabulary

Round, approximately

Resources

Number lines; mini-whiteboards; digit cards 0–9 or 10-sided dice

Background knowledge

The rules of rounding: if a number ends in 5, 6, 7, 8 or 9 then round up; if the number ends in 0, 1, 2, 3 or 4 then round down. This is easier to see on a number line, but children sometimes think that rounding down means the relevant digit decreases by 1 instead of staying the same. The other common difficulty is rounding larger numbers. Children might successfully round to the nearest 100, but may not write all the digits of the original number; for example, when rounding 34 567 to the nearest 100 they may just write 600, rather than 34 600.

Teaching activity 1a (15 minutes)

Round four-digit, five-digit and six-digit numbers to the nearest 1000, 10 000 and 100 000 respectively

- Recap the rules of rounding, emphasising that mathematicians have decided that 5, 50 and 500 will all round up.
- Ask children to suggest (or use digit cards to generate) a four-digit number, such as 6398, to round to the nearest 1000. Write the number on a whiteboard. Ask: **Which are the two possible numbers it could round to?** *(6000 and 7000)* **Which digit do we need to look at to decide?** *(The hundreds digit)* **Is the hundreds digit more or less than 5?** *(It is 3, so it is less than 5, so it rounds down to 6000, meaning the thousands digit stays the same.)*
- Repeat, using a different number, asking children to supply the two possible answers before deciding on the correct one.
- Move on to rounding five-digit numbers to the nearest 10 000. Ask: **Which digit is the important one now?** *(The thousands digit)* Ask the children to round 37 129 to the nearest 10 000. Ask: **Which are the two possible numbers it could round to?** *(30 000 and 40 000)* **Which digit must we look at?** *(The 7, and as it is above 5 the number rounds up to 40 000.)*

- Repeat several times, manipulating the numbers so that the children practise rounding up and down.
- Repeat the activity, using six-digit numbers and rounding to 100 000. Children now need to consider the second digit, the tens of thousands. Continue to ask for the two possible answers before deciding.

Teaching activity 1b (15 minutes)

Round four-digit, five-digit and six-digit numbers to the nearest 1000, 10 000 and 100 000 respectively

- Draw a number line on a whiteboard, labelling the ends _000 and _000 and the middle _500. Choose or generate a four-digit number, such as 8395. Explain that this number is on the number line somewhere. Ask: **What numbers should we write at each end and in the middle on the number line so that we can round 8395 to the nearest 1000?** Establish that the numbers should be 8000, 8500 and 9000. Ask children to suggest where, approximately, 8395 would be on this line. Ask: **Is it more or less than 8500?** *(less)* **Is it nearer to 8000 or 8500?** *(nearer to 8500)* Establish that it is less than 8500 and more than halfway between 8000 and 8500.

Explain that this shows that 8395 is closer to 8000 than to 9000, so it rounds to 8000. Repeat with a few more four-digit numbers.

- Repeat with five-digit numbers, rounding to nearest 10000. Label the ends of the number line _0000 and _0000 and the middle _5000. Ascertain that, in a number that rounds up, the last four digits show a number greater than or equal to 5000.
- Repeat with six-digit numbers, rounding to nearest 100000. Label the ends of the number line _00000 and _00000 and the middle _50000. Ascertain that in a number that rounds up, the last five digits show a number more than 50000.

Teaching activity 2a (15 minutes)

Round any number to the nearest 10, 100, 1000, 10000 and 100000

- Start with a few four-digit numbers, rounding each to the nearest 10, 100 and 1000. Avoid using the digit 9 at this stage. Ensure that children realise that if they start with a four-digit number, then the answer should also have four digits and the number of zeros should match the number they are rounding to.
- Now introduce numbers in which the digit to be rounded is 9, such as 5397, 4982, 5938, and show what happens. As a final number, choose a number such as 9807, which rounds up to 9810, down to 9800, then up to a five-digit number, 10000, so it will have one more zero than 1000. Ask: **What has happened?** (*When we round up, 9 thousands become 10 thousands.*) Explain that a four-digit number will only round up to a five-digit number when the thousands digits is 9 and the hundreds digit is greater than or equal to 5.
- Continue with five-digit numbers, ensuring that the rounded answer is a five-digit number. Include some numbers in which the important digit is 9, such as 45912, which rounded to the nearest 100 becomes 46000, and 38296, which rounded to the nearest 10 becomes 38300. Finish with an example such as 95450 rounded to the nearest 10000, which will give the six-digit number 100000.
- Continue with rounding six-digit numbers, ensuring that the rounded number also has six digits. Again, finish with an example such as 972399 to the nearest 100000, which will round to the seven-digit number 1000000.

Teaching activity 2b (15 minutes)

Round any number to the nearest 10, 100, 1000, 10000 and 100000

- Draw an empty number line on a whiteboard. Start with five-digit numbers, generated from digit cards or dice. Round to the nearest 1000, then 100, then 10, by writing the relevant numbers on the ends and in the middle of the line so that children can see why a number rounds up or down. To round 24563 to the nearest 100, label the number line with 24500, 24550 and 24600. Ask: **Approximately where will 24563 be on this line?** (*After 24550, so it rounds up to 24600.*) **Who can suggest a number that will round down to 24500? Another number that will round up to 24600?**
- Now draw a number line labelled 4560, 4565 and 4570 and tell children they are rounding to the nearest 10. Ask for suggestions for numbers that will round up to 4570 or down to 4560. Ask: **Can you list all the numbers from 4560 to 4570 that will round up to 4570? Now list the numbers that will round down to 4560.**
- Repeat, this time rounding six-digit numbers to the nearest 10, 100, 1000, 10000 and 100000, then asking children to suggest other numbers that will round up or down, as above.

Unit 5: Read Roman numerals to 1000 (M) and recognise years written in Roman numerals

Content domain reference: 5N3b

Prerequisites for learning

Be aware of Roman numerals I, V, X, L and C

Key vocabulary

Roman numerals

Learning outcomes

Understand how Roman numerals are formed

Read Roman numerals to 1000 and recognise dates written in Roman numerals

Resources

Resource 4: Roman numerals 1–100; Resource 5: Roman numeral cards (a set of cards for each child)

Background knowledge

Roman numerals have been around for over 2000 years. They were used to record numbers in text carved in stone, in art and on coins throughout the Roman Empire. They are still used today, for example, in copyright dates, to mark film sequels and in some books for chapter headings and in lists. Some clock faces use Roman numerals. The system follows certain rules, which are a good starting point for the lesson.

Once the children know the symbols, encourage them to learn the rules, to read and then write them.

1 Roman numerals are written in order of place value, starting with the largest.

CCLXVIII is $100 + 100 + 50 + 10 + 5 + 3 = 268$.

2 When a symbol with a smaller value is written before a symbol with a larger value, subtract it. These all involve the decimal digit 4 or 9, but there are other rules that govern what you can and cannot write in this way.

 • I can be subtracted from V and X but not from L, C, D or M.

 IV means $5 - 1 = 4$, IX means $10 - 1 = 9$

 • X can be subtracted from L and C but not from D or M.

 XL means $50 - 10 = 40$, XC means $100 - 10 = 90$

 • C can be subtracted from D or M.

 CD means $500 - 100 = 400$, CM means $1000 - 100 = 900$

3 It is never necessary to repeat the symbol V, L or D. A double gives a total that has its own letter.

 VV is $5 + 5 = 10$ (X), LL is $50 + 50 = 100$ (C), DD is $500 + 500 = 1000$ (M).

4 The symbols I, X, C and M can be used a maximum of **3** times:

 II means 2, XXX means 30, CC means 200, MMM means 3000.

These rules and the Roman equivalents for numbers 1–100 are all on Resource 4: Roman numerals 1–100 and will take you as far as 3999.

Teaching activity 1a (15 minutes)

Understand how Roman numerals are formed

• Ask children what they remember about Roman numerals. Ask: **Where have you seen Roman numerals? Can you remember any of the letters? The rules?** They should have covered the first five letters: I, V, X, L and C. Explain that there are seven Roman symbols to recognise. Revise what each letter is worth, adding the extra letters D and M: 1, 5, 10, 50, 100, 500, 1000. Ask: **What do you notice?** (*They all involve 1 or 5.*)

• Go through the rules for using Roman numerals, following each rule in turn to form numbers.

• Practise forming two-digit numbers as revision, referring to the rules each time and checking answers against Resource 4: Roman numerals 1–100.

Teaching activity 1b (15 minutes)

Understand how Roman numerals are formed

- Make up a mnemonic to remember which letter is which. Establish that the equivalents in our number system are just the numbers 1, 5, 10, 50, 100, 500 and 1000 and all other numbers are formed in Roman numerals with combinations of the these letters, following certain rules. Here are some possibilities.
  - It's **V**ery e**X**citing **L**earning **C**ollin's **D**aily **M**aths
  - **I V**alue **X**-boxes **L**ike **C**ats **D**o **M**ilk
- If possible, make a mnemonic for your group with some significant words that they will remember, or learn one of these.
- Give each child a set of cards with the Roman symbols and their equivalents, cut from Resource 5: Roman numeral cards. Hold up 1, 5, 10, 50, 100, 500 or 1000 and ask the children to show you the corresponding Roman numeral. Repeat, this time showing the children a Roman numeral and asking the children to show you the corresponding number.

Teaching activity 2a (15 minutes)

Read Roman numerals to 1000 and recognise dates written in Roman numerals

- Together, checking the rules carefully, work out some larger numbers given in Roman numerals, for example:
  - MCXXV: 1000 + 100 + 20 + 5 = 1125
  - MMDCCCXLIII: 2000 + 500 + 300 + 40 + 3 = 2843.
- Together, form some Roman numerals from two-digit numbers suggested by the children, checking against Resource 4: Roman numerals. When the children are confident with this, move on to three-digit numbers, then four-digit numbers, showing some examples first. Ensure that you include 4s and 9s in some numbers, for example:
  - 378: 300 + 50 + 20 + 5 + 3 = CCCLXXVIII
  - 2469: 2000 + (500 – 100) + 50 + 10 + (10 – 1) = MMCDLXIX
- Give children, in pairs, a set of numbers to translate into Roman numerals.
- Set them to work out some key dates from a recent or current history topic. 2018: MMXVIII; 1066: MLXVI; 1611: MDCXI; 1215: MCCXV; 1559: MDLIX; 1896: MDCCCXCVI; 1914–1918: MCMXIV–MCMXVIII
- Ask children to write some numbers in Roman numerals. Ask them to swap with a partner, who checks they have kept to the rules and then translates the numbers.

Teaching activity 2b (15 minutes)

Read Roman numerals to 1000 and recognise dates written in Roman numerals

- Children use the cards on Resource 5: Roman numeral cards to form the numbers 1–20, revising how the pattern builds up, and the rules. They form the multiples of 10 for 30–90, remembering that 50 has its own symbol, L, so 40 = XL; then they form the multiples of 100: 100–900, remembering that 500 has its own symbol, D, so 400 = CD and 900 = CM because 1000 = M.
- Invite children to take turns using the Resource 5: Roman numeral cards to form combinations of Roman numerals to translate into ordinary numbers. Refer back to the rules each time.
- Work out some key dates or dates from a recent/current history topic. 2018: MMXVIII; 1066: MLXVI; 1611: MDCXI; 1215: MCCXV; 1559: MDLIX; 1896: MDCCCXCVI; 1914–1918: MCMXIV–MCMXVIII

Unit 6: Solve number problems and practical problems that involve objectives in units 1–5

Content domain reference: 5N6

Prerequisites for learning

Read, write and compare numbers up to 1 000 000

Round numbers to 10, 100, 1000, 10 000 or 100 000

Read Roman numerals to 3000 (MMM)

Order and interpret negative numbers in context

Count forwards or backwards in powers of 10

Interpret word problems involving numbers to 10 000

Learning outcomes

Solve problems relating to place value involving numbers to 1 million, negative numbers and Roman numerals

Key vocabulary

Population, temperature, thermometers, increase, decrease, average, approximately, census, order, negative, minus, degrees (see units 1–5)

Resources

Mini-whiteboards; digit cards 0–9; slips of coloured paper (a different colour for each category), one for context, one for the operation and one for the number as follows:

Context: temperature, Roman numerals, population, census, degrees, whole number, negative

Operation: order, round, increase, decrease, approximate, minus, numerals, four-digit, five-digit, six-digit

Numbers: 10, 100, 1000, 10 000, 100 000

Background knowledge

These word and practical problems involve the concepts covered in units 1–5. It can be helpful for children, using the starting point of key words and key numbers, to write their own word problems to solve so that they think about the words they use.

Teaching activity 1a (20 minutes)

Solve problems relating to place value involving numbers to 1 million, negative numbers and Roman numerals

- With the children, write a list of key words and numbers they would expect to see in a problem relating to:

  - reading, writing, ordering whole numbers (order, compare, numerals, digit, value)

  - rounding whole numbers (round, approximately, estimate)

  - temperature (temperature, increase, decrease, lower, higher, degrees, minus)

  - Roman numerals (translate, numeral)

  - population and census figures (census, population, increase, decrease, approximate).

- Show children the words written on the slips of coloured paper. Ask: **Which words/numbers would you expect in a word problem involving rounding, ordering negative numbers, temperature, Roman numerals, population figures? Which words could be used in more than one/all of these?**

Teaching activity 1b (20 minutes)

Solve problems relating to place value involving numbers to 1 million, negative numbers and Roman numerals

• Use the context, operation and number on the slips of coloured paper to help children to write some word problems. Choose a context card first, then decide if you need an operation and/ or a number/digit card to generate the number. (The context card may be enough to trigger a question.)

• If an unhelpful combination is drawn, such as negative/Roman numeral – approximate/round, choose a different slip, write a problem based on just one slip or draw a third and identify the best match of two cards.

• Here are some starter examples.

  ◆ Temperature – increase – 6: **The temperature increased** by **6** degrees to 4 degrees centigrade. What was the starting temperature?

  ◆ Population – rounding – 10 000: The **population** of a town is 28 490. What is the population **to the nearest 10 000?**

  ◆ Roman numeral – four-digit – 3284: Write the number **3284** in **Roman numerals**.

  ◆ Roman numeral – order: **Order** these dates by writing them in ordinary numerals.

  ◆ Census – approximate – 1000: The **census** says that a town has 26 809 people living in it. **Approximately** how many **thousand** people is that?

  ◆ Whole number – order: Here are some lengths of rivers, in **whole** kilometres. **Order** the rivers, longest to shortest.

• Ask children, in pairs, to write their own word problems on a piece of paper. They can use the context slips along with optional operations/ number slips. They should work out the answer, writing it on the back, to ensure that it is possible to solve the problem and that it makes sense. They swap with another pair to work out the answers, checking with the answer on the back. Ask: **Do you agree with the answer given?** If there is disagreement, work through the problem with them all.

• Ask children to write word problems without referring to the coloured slips but using the same concepts. Other children attempt to work out the answer. Ask: **Is this a good question? Can you answer this question with the skills you have practised? Is there any information in the question that you don't actually need? Which are the important words and numbers in this question? What do you have to do to answer this question? Is it a one-step or a two-step question?**

• If it is helpful, categorise the questions on the practice page. Ask: **What are the key words? What is this question asking us to do?**

Unit 7: Add and subtract whole numbers with more than four digits, including using formal written methods (columnar addition and subtraction)

Content domain reference: 5C2

Prerequisites for learning

Add and subtract whole numbers with four or more digits, including using a formal written method

Learning outcomes

Use a formal written column method to add numbers

Use a formal written column method to subtract numbers

Key vocabulary

Sum, difference, column method, exchange

Resources

Digit cards (optional); 1 cm squared paper; Dienes apparatus

Background knowledge

Examiners are increasingly insistent that children use traditional formal written methods for all four operations. The decision as to where to write carried numbers in addition and how to record exchanges across the columns in subtraction will be a whole-school policy. This lesson shows the most common approach. In SATs the question may be presented horizontally, but the expectation is that children rewrite it vertically in a column style.

Teaching activity 1a (15 minutes)

Use a formal written column method to add numbers

- Write an addition, such as 4967 + 2846, horizontally on a whiteboard. Ask: **What type of calculation is this?** (*Addition*) **How can we work this out? Can we do this calculation in our heads?** (*It would be hard because they are four-digit numbers.*) Explain that this type of addition really needs a written method.

- Demonstrate how to rewrite it in columns. Work through it, writing the carried digits under the line, aligned to the correct column and talking through it as you work: **7 + 6 is 13, 1 ten and 3 ones, so I write the 3 here in the ones column and I write the 1 ten under the tens column, here. Now the tens: 6 + 4 + the carried 1 = 11,** etc. **Why do we have to start with the smallest digits, the ones?**

```
    4   9   6   7
+   2   8   4   6
─────────────────
    7   8   1   3
    1   1   1
```

- Repeat with two five-digit numbers, then two six-digit numbers, explaining that the method stays the same. You may wish to demonstrate addition of three or more numbers of the same length to show that the method is still exactly the same.

- Now add a five-digit number to a four-digit number. Write it incorrectly, so that the digits are not lined up. Ask: **Is this right? What have I done wrong?** (*You haven't put the ones, tens, hundreds,… digits in the correct column.*) **Can someone show me how to write it correctly? Why is it important to line up the digits?** (*So you can add them correctly. So you are not adding a ones digit to a tens digit.*) **Which is the hundreds digit in this number? And in this number?** Say: **The digits with the same place value must be directly under each other.** Work out the answer and repeat.

- Finally, add together several numbers of different lengths: 47 235 + 509 + 3 296. Ask children to write the sum on their mini-whiteboards and then let them check that their partner has written it correctly. They then work it out and check answers.

Teaching activity 1b (15 minutes)

Use a formal written column method to add numbers

- Write the place-value headings on centimetre-squared paper, Th, H, T, O, one heading per square across the paper. Write an addition calculation such as 4967 + 2846. Talk through how to work it out by the column method, writing each digit under the correct place-value heading.

- Work out each column, starting with the ones and using Dienes apparatus, where helpful, to show what happens when the total of a column is more than 9. Record the carried number under the line in the relevant column.

- Discuss the carried numbers, demonstrating that 13 is 1 ten and 3 ones. Say: **So we write the 3 in the ones column and the ten in the tens column.** You may need to explain that a two-digit total is written in the same order as it usually is, so the first digit is always the carried number and the second digit is written in the column, between the lines.

- Now write 3587 + 21 + 376. Talk through how to set out this sum in columns, asking key questions: **Which digit has the largest value in this number? Where do we need to write that digit? Which digit is the hundreds in this number?** Complete the calculation, using Dienes where helpful.

Teaching activity 2a (15 minutes)

Use a formal written column method to subtract numbers

- Write a subtraction, such as 6945 – 2856, on a whiteboard (horizontally). Ask: **What type of calculation is this?** *(subtraction)* **How can we work this out? Can we do this calculation is our heads?** *(You can but it will be difficult, so doing a written calculation would be easier.)* Explain that this type of subtraction really needs a written method.

- As with addition, write it out vertically, matching the digits carefully and talking through the method: **In the ones column, we have 5 – 6. We cannot do this, so we exchange 1 ten for 10 ones, leaving 3 tens. Now we have 15 – 6 = 9. Now, in the tens column, we have 3 – 5. We cannot do this, so exchange 1 hundred for 10 tens, leaving 8 hundreds. Now we have 13 – 5 = 8. In the hundreds column we have 8 – 8 = 0, so we do not need to exchange this time. Finally, for the thousands, 6 – 2 = 4.**

$$
\begin{array}{r}
6 \;\; {}^{9}\!\!\!\not{9}{}^{8} \;\; {}^{1}\!\!\!\not{4}{}^{3} \;\; {}^{1}5 \\
- \;\; 2 \quad 8 \quad 5 \quad 6 \\
\hline
4 \quad 0 \quad 8 \quad 9
\end{array}
$$

- Repeat with two five-digit numbers and then two six-digit numbers, ensuring that exchanging will be required for at least two digits in the larger number, for example: 38 217 – 24 672 and 826 491 – 453 855.

- Repeat with some zeros in the top line, where the exchange has to go over two columns, one at a time; for example: 90 004 – 34 728. Record the exchanged digit as 10 in the next column initially, then exchange again, leaving 9 in the 'zero' column(s).

$$
\begin{array}{r}
9 \quad 0 \quad 0 \quad 0 \quad 4 \\
{}^{8}\!\!\!\not{9} \;\; {}^{1}0 \quad 0 \quad 0 \quad 4 \\
{}^{8}\!\!\!\not{9} \;\; {}^{1}\!\!\!\not{0}{}^{9} \;\; {}^{1}0 \quad 0 \quad 4 \\
{}^{8}\!\!\!\not{9} \;\; {}^{1}\!\!\!\not{0}{}^{9} \;\; {}^{1}\!\!\!\not{0}{}^{9} \;\; {}^{1}0 \quad 4 \\
{}^{8}\!\!\!\not{9} \;\; {}^{1}\!\!\!\not{0}{}^{9} \;\; {}^{1}\!\!\!\not{0}{}^{9} \;\; {}^{1}\!\!\!\not{0}{}^{9} \;\; {}^{1}4 \\
- \;\; 5 \quad 5 \quad 2 \quad 7 \quad 6 \\
\hline
3 \quad 4 \quad 7 \quad 2 \quad 8
\end{array}
$$

- Repeat, subtracting a number with fewer digits, such as 76 254 – 8460, talking through how to line up the digits correctly and why this is important, as in addition.

Teaching activity 2b (15 minutes)

Use a formal written column method to subtract numbers

- Use Dienes apparatus (or number cards) with two four-digit numbers to demonstrate what happens when a number is exchanged. If using number cards, replace the relevant digit with 'one less' and use '1' as the number carried over to show a two-digit teens number in that column.

- Explain that the Dienes apparatus does not go far enough for larger numbers, so continue using squared paper with headings or digit cards, ensuring that children understand what to do when there are zeros in the top line. A common error is to jump over any column with a zero in it.

Unit 8: Add and subtract numbers mentally with increasingly large numbers

Content domain reference: 5C1

Prerequisites for learning

Add and subtract two-digit numbers mentally
Know complements to 100

Learning outcomes

Add and subtract mentally, multiples and near multiples of 100, 1000 and 10000
Add and subtract larger numbers using knowledge of smaller numbers

Key vocabulary

Complements to 1000; adjust; near multiple

Resources

Resource 6: 100 square (one per child and one enlarged to A3); digit cards or dice to generate numbers

Background knowledge

Mathematics is all about connections. For example, if you know 35 + 45 = 80, then you also know that 350 + 450 = 800. This lesson concentrates on using what children know about two-digit numbers, to work with larger values and to show that it can be more efficient to add or subtract mentally than to use a written method.

Teaching activity 1a (15 minutes)

Add and subtract mentally, multiples and near multiples of 100, 1000 and 10000

- Remind children how they used place value to add and subtract 100, 1000 and 10000 in Unit 2. Revisit if necessary. Explain that now they can use the same method to add multiples of 100, 1000 and 10000.

- Count on and back from a chosen four-digit number in steps of 200 and 500; repeat in steps of 3000 and 5000 from a four-digit number, then from a five-digit number; repeat in steps of 20000 and 50000 from a five-digit number, then from a six-digit number.

- Generate a four-digit number and ask a child to suggest what to add or subtract (*a multiple of 100 or 1000*); repeat with a five-digit number, then a six-digit number, adding or subtracting a multiple of 100, 1000 or 10000. Record each calculation as an addition or subtraction, each time considering which digit has changed.

- Write 5423 + 2999 and ask for suggestions for how to do this mentally. Explain that 2999 is a near multiple of 1000: it is very close to 3000. Ask: **How could we use this?** (*Add 3000 then subtract the extra 1: 5423 + 3000 = 8423, 8423 − 1 = 8422.*) Repeat with other four-digit numbers, adding a near multiple of 1000 each time.

- Repeat for subtraction, starting with 5423 − 2999. Ask: **Can we use the same method as for addition? How do we need to adjust the answer?** (*Add one, because we have subtracted one too many. 5423 − 3000 = 2423, 2423 + 1 = 2424.*) Repeat with other four-digit numbers, subtracting a near multiple of 1000 each time.

- If appropriate, repeat for addition and subtraction with near multiples of 10000.

Teaching activity 1b (15 minutes)

Add and subtract mentally, multiples and near multiples of 100, 1000 and 10000

- Use digit cards to demonstrate adding multiples of 100 to four-digit, five-digit then six-digit numbers by physically exchanging the hundreds cards. Include at least one example in which the hundreds will add to more than 9, to show that the thousand digit will also change: 3768 + 400 = 4168 (7 + 4 = 11, 3 + 1 = 4).

- Repeat for subtraction.

- Repeat for addition and subtraction of multiples of 1000 then 10000.

- Write 5423 + 2999 and ask for suggestions on how to do this mentally. Use digit cards to demonstrate, explaining that 2999 is very close to 3000. Add the 3000 as before. Ask: **Have I added too much or not enough?** *(too much)* **What can I do to get the right answer?** *(subtract 1)* Exchange the ones cards to show the adjustment of – 1.

- Repeat for subtraction, exchanging the ones card to show the adjustment of + 1.

Teaching activity 2a (15 minutes)

Add and subtract larger numbers, using knowledge of smaller numbers

- Start with quick-fire complements to 100. For example, write a two-digit number on a whiteboard, ask the children to write its complement to 100: you write 45, they write 55. Check their answers, as a common error is ending up with a total 10 too high. Explain that unless both numbers end in zero, the ones must add to 10 but the tens must add to 9, as the extra 10 comes from the ones. Record some of these as both additions and subtractions: 32 + 66 = 100; 100 – 71 = 29.

- Repeat with three-digit multiples of 10, asking for complements to 1000 such as 350 and 650.

- Repeat with four-digit multiples of 100, asking for complements to 10 000 such as 3500 and 6500.

- Now try other three-digit multiples of 10 and four-digit multiples of 100, explaining that you only need to add or subtract the two digits at the beginning of the number and adjust: 220 + 450 = 670 because 22 + 45 = 67; 34 000 – 12 000 = 22 000 because 34 – 12 = 22; 450 – 210 = 240 because 45 – 21 = 24; 56 000 – 35 000 = 21 000 because 56 – 35 = 21.

Teaching activity 2b (15 minutes)

Add and subtract larger numbers, using knowledge of smaller numbers

- Show children Resource 6: 100 square and how to use it to work out all the number pairs to 100. Choose a number, then move down the column, using the rows to count in tens to the bottom of the 100 square. Then count along the bottom row to find how many ones to get to 100. Work out and record some pairs: 36 + 64 = 100; 100 – 64 = 36.

- Explain that they can use this to find pairs to 1000 and even 10 000. Consider 250: start on 25 but write a zero on it to make it 250. Count down the columns to see how many hundreds, then count across the bottom row to find how many tens to get 700 + 50 = 750.

- Some children will be able to see straight away that if 25 + 75 = 100, then 250 + 750 = 1000 without counting. This is what you want them to understand.

- Now try other three-digit multiples of 10, counting on for addition and back for subtraction, in 100s then 10s, always starting with the larger number: 220 + 450: 450, 550, 650, 660, **670**; 450 – 210: 450, 350, 250, **240.**

- If appropriate, repeat for four-digit multiples of 100, counting in 1000s then 100s.

Unit 9: Use rounding to check answers to calculations and determine, in the context of a problem, levels of accuracy

Content domain reference: 5C3

Prerequisites for learning

Be able to round any number to the nearest 10, 100, 1000, 10000

Add and subtract two-digit numbers mentally

Learning outcomes

Use rounding to estimate and check answers to calculations

Key vocabulary

Estimate, approximate, ≈, check, accurate, error

Resources

Mini-whiteboards; calculators (optional)

Background knowledge

Children very seldom check their answers and therefore make avoidable errors. Encourage them always to check each answer to ensure that it makes sense. Rounding is a useful way to estimate answers, but it rarely gives the actual answer. Children can often be confused at what they see as quite a difference between the estimation and the actual answer. They tend to want to round to the nearest 10 or 100, but this leaves them with difficult numbers to calculate mentally. Rounding to the nearest thousand, although less accurate, gives numbers that are much easier to manipulate mentally. Children can use calculators or written methods to check answers against the estimates.

Teaching activity 1a (20 minutes)

Use rounding to estimate and check answers to calculations

• Discuss why checking answers is so important. Then show children some miscalculations to check whether they can spot the errors that have been made.

♦ **Addition/subtraction:** 560 – 120 = 680 (*added instead of subtracted*); 890 + 150 = 740 (*subtracted instead of added*)

Empty boxes to fill in: $\boxed{360}$ – 120 = 480 (*added instead of subtracted*); $\boxed{240}$ + 40 = 200 (*subtracted instead of added*)

• Recap on the rules of rounding.

• Write 6285 + 3876 on a whiteboard and demonstrate how to use rounding to find an approximate answer, beginning with rounding to 10: 6290 + 3880. Ask: **Is this easy to work out in our heads?** (*No, we may as well work out the actual accurate answer, so rounding to 10 does not help very much.*)

• Suggest rounding to 100s: 6300 + 3900. Ask: **Can we work this out easily in our heads?** (*approximate answer 10200*).

• Show that rounding to 1000s is the easiest to do mentally: 6000 + 4000 = 10000.

• Use a calculator or column method to work out the actual answer (10161). Ask: **Which method gives the more accurate answer?** (*rounding to 100s*) **Which was the easier to work out in our heads?** (*rounding to 1000s*)

• Establish that an estimation needs to be done quickly and mentally, so although it may be less accurate, it is best to round to a higher place value. Depending on the actual numbers, it can also be straightforward to round to the second highest place value.

• Repeat with 7842 – 2855. Say: **Rounding to 10 gives 7840 – 2860.** (*too difficult to work out quickly*) **Rounding to 100s gives 7800 – 2900 = 4900.** (*possible but still a bit awkward*) **Rounding to 1000s gives 8000 – 3000 = 5000.** (*quick and easy and not very far from 4900*) The actual answer will be less than 5000 since, although both numbers will be rounded up, 842 < 855. Work out the actual answer and compare with the estimates.

- Repeat with some five-digit numbers: 32 146 + 17 849. Rounding to 100s gives 32 100 + 17 800 = 49 900; rounding to 1000s: 32 000 + 18 000 = 50,000; rounding to 10 000s: 30 000 + 20 000 = 50 000. Establish that in this example rounding to both 1000s or 10 000s is quite easy and, interestingly, they both give the same answer. Work out the actual answer and compare with estimates.

Teaching activity 1b (20 minutes)

Use rounding to estimate and check answers to calculations

- Ask children to write all the hundreds bonds to 1000. Encourage a systematic approach (100 + 900, 200 + 800, …).

- Write 1000 at the top of a whiteboard and explain that, when two numbers are added together, this is the approximate total. Ask a child to choose one of the number bonds, such as 600 + 400, and write it underneath. Ask: **Who can give me a number that will round** up **to 600?** (a number from 550 to 599) **Who can give me a number that will round** up **to 400?** (a number from 350 to 399) Choose one of each and write the calculation underneath; repeat with another pair of numbers, for example:

```
    1000           1000
 600 + 400      600 + 400
 586 + 359      591 + 350
```

- Say: **We have done some rounding in reverse. 586 + 359 and 591 + 350 are both approximately 1000.** Introduce the symbol ≈, saying: **We can use this symbol to show that an answer is approximate.** Write 586 + 359 ≈ 1000. Ask: **Will this actual answer be more or less than 1000?** (Less, as both numbers were rounded up.)

- Repeat, but this time use a different bond (300 + 700) and identify numbers that will round **down** to 300 (numbers from 301 to 349) and 700 (numbers from 710 to 749). Ask: **Will the actual answer be more or less than 1000?** (More, as both numbers were rounded down.)

- Ask children quickly to note down the thousands bonds to 10 000 (1000 + 9000, …) and choose one as a starting point, such as 2000 + 8000. Repeat the above steps, choosing numbers that round to 2000 and 8000. Ask: **Can we tell if the answer is going to be more or less than our estimate?** (Yes if both numbers have been rounded up or down; not easily if one is rounded down and the other rounded up, but by comparing the HTOs digits it is quite easy to work it out.) Finally, choose two multiples of 1000, then 10 000, such as 2000 + 14 000, to make a subtraction. Work out the answer, choosing pairs of numbers that round to the given thousands, as before, then work backwards up to the estimate, for example:

```
8000 − 6000 = 2000       7607 − 5990 ≈ 2000
18 000 − 4000 = 14 000   17 945 − 4123 ≈ 14 000
```

Unit 10: Solve addition and subtraction multi-step problems in contexts, deciding which operations and methods to use and why

Content domain reference: 5C4

Prerequisites for learning

Add and subtract whole numbers with more than four digits

Key vocabulary

Total, sum, difference, how many more than, how many less than, fewer, change, left over

Learning outcomes

Solve multi-step problems involving addition and subtraction

Solve number puzzles

Resources

Digit cards (one set of eight per pair); calculators (optional)

Background knowledge

The best approach to answering single- and multi-step problems is to identify the key words and numbers and to record each step, looking back at the question to make sure you have answered it correctly and not missed something. Often, children answer the first part but do not complete the question.

Teaching activity 1a (15 minutes)

Solve multi-step problems involving addition and subtraction

- Make a list of words associated with addition, then another for subtraction. Say: **We are going to make up some word problems to solve together.** Start with this one as an example: **Town A has a population of 32 856; Town B has a population of 18 923. Identify the key words and numbers by underlining them. What is the** difference **in their population?** Ask: **Is this an addition or a subtraction problem? How many more people live in town A? What is the** total **population of both towns?**

- Add Town C with a population of 13 096. Ask: **Do more or fewer people live in Town A than in Town B and C combined? What do we need to do first to solve this problem? What is the next step? Is this an addition or a subtraction problem, or both?**

- Emphasise that when a question says 'show your working' it means that you should write down each calculation, in the order you do them, even if you can actually do the calculation mentally. Ask: **Have you answered the question? Do you need to work out how many fewer/more?**

- With the children, write and solve some word problems that involve both addition and subtraction, giving a starting context if you need to, such as seats at a sports stadium, population, cost of a house or car. Highlight key words and numbers, identify operation(s). Ensure that there is more than one step involved in solving the problem.

Teaching activity 1b (15 minutes)

Solve multi-step problems involving addition and subtraction

- Start by writing 24, 76 and 100 on a whiteboard. Explain that two of these numbers are the question and one is the answer. Ask: **What could the question be? How many different ways could these three numbers be combined to make a question and answer?** *(24 + 76 = 100; 76 + 24 = 100; 100 – 24 = 76 and 100 – 76 = 24)*

- Write 365, 189 and 176 on a whiteboard. Explain that the question has two numbers in it, 365 and 189 and the answer to the question, 176. Ask: **What type of question must it be, addition or subtraction? How do you know?** *(Subtraction, as 176 is smaller than either number in the question.)* Write 365 – 189 and 189 – 365 on a whiteboard. Ask: **Which is the correct subtraction for the answer 176?** *(365 – 189 because the larger number is first.)*

- Point to the numbers on the whiteboard and ask: **Can we write a word problem to match this calculation?** *(There have been 189 days in this year, so far. How many days are left?)* **What do we need to know to answer this question?** *(365 days = year)* **What else could we ask, to make it multi-step?**

- Suggest trying to find the actual date, or at least the month, when it will be/was day 189 of the current year. Start with an estimate. Ask: **Is this less or more than halfway through the year?** *(Half of 365 is $182\frac{1}{2}$ so day 189 is a little more than halfway.)* Say: **Work out the total number of days for the first 6 months: 31 + 28 (29 if a leap year) + 31 + 30 + 31 + 30 = 181 or 182 (end of June) so 8 or 9 July.**

- Ask the children, in pairs, to make up an **addition** word problem involving the same three numbers. Ask: **Which of the numbers must be the answer?** *(365)* Check each pair's ideas to make sure they have written an addition problem, then, using more numbers if necessary, make it into a multi-step problem.

- Repeat with three, four-digit numbers (or two-digit numbers if necessary) in the question, trying to ensure that the children write at least a two-step question. Write some examples of what the calculation could be: 4756 + 2665 – 1295; 4756 – 2665 + 1295; 1295 + 4756 – 2665. Encourage the children to add extra parts to the question to give it as many steps as possible, using more numbers if necessary, but still only involving addition and subtraction. Solve each pair's problem, recording each step carefully. If appropriate, use estimation to check that the question is likely to work.

Teaching activity 2a (15 minutes)

Solve number puzzles

- Use digit cards to make a four-digit + four-digit addition and a four-digit – four-digit subtraction. Work out the answer and use digit cards to show the answer.

- Remove the first and third digit of the top number and the second and fourth digit of the second number in the addition. Ask: **How could we work out these missing digits if we did not know what they were?** Repeat with the subtraction.

- Repeat with different calculations, taking care to include examples in which there are hidden carried digits in the addition (columns adding to more than 10) and hidden exchanges in the subtraction (lower-number digit higher than the upper-number related digit).

Teaching activity 2b (15 minutes)

Solve number puzzles

- Pick out eight digit cards and give each pair a set of these cards. Explain that they must use these digits to make the largest possible total and the smallest possible difference. Let children arrange the cards to make a four-digit + four-digit addition.

- Compare each pair's additions. Ask: **Who has the largest total? Is there an even larger total?** Establish that the highest two numbers on their cards must be used for the two thousands digits, the next highest the hundreds, and so on.

- Repeat with subtraction, establishing that, the digits used for the thousands must have the smallest difference (consecutive numbers give this), then the digits for hundreds, tens and units. They need to choose small digits for the top number, larger digits for the bottom number.

Unit 11: Identify multiples and factors, including finding all factor pairs of a number, and common factors of two numbers

Content domain reference: 5C5a

Prerequisites for learning

Knowledge of multiplication tables to 12 × 12
Know division facts related to multiplication facts

Key vocabulary

Factor, multiple, factor pair, common factor, divisible by

Learning outcomes

Identify multiples of numbers 1–12
Identify factors, factor pairs and common factors of two-digit numbers

Resources

Resource 7: Multiplication grid; counters (at least 24 per pair)

Background knowledge

A **multiple** of a number appears in its multiplication table, for example, 24 is a multiple of 1, 2, 3, 4, 6, 8, 12 and 24. Every positive number is a multiple of 1. This introduction focuses on multiples up to 12×. A **factor** of a number divides exactly into that number with no remainder: 1, 2, 3, 4, 6, 8, 12 and 24 are factors of 24; **1 and the number itself** are always factors of that number. A **factor pair** of a number comprises two numbers that multiply together to make that number: 2 and 12, 3 and 8, 4 and 6 are all factor pairs of 24. A **common factor** of two numbers is a number that is a factor of both numbers: 3 is a factor of 24 and it is a factor of 15, so 3 is a common factor of 24 and 15.

Teaching activity 1a (15 minutes)

Identify multiples of numbers 1–12

- Count on and back from zero in multiples of 2, 5 and 10. Explain that these numbers are **multiples** of 2 (or 5, 10) because they are in the 2× (or 5×, 10×) tables.

- Chant a times table they need to practise (usually 6, 7 and 8), writing the numbers as a sequence on the board as you say them: **1 times 6 is 6; 2 times 6 is 12... 12 times 6 is 72.** Say: **The numbers I have written are all *multiples* of 6. The first multiple of 6 is 6, the second multiple of 6 is 12, what is the fifth multiple of 6?** (30) **And the 8th multiple?** (48)

- Repeat with a different times table, asking similar questions as before. Then ask: **How could we work out the seventh multiple of 10 without counting?** (7 × 10) **The fourth multiple of 9?** (4 × 9) **The third multiple of 12?** (3 × 12)

- Ask some quickfire questions: **What is the sixth multiple of 3? The tenth multiple of 5?**

Teaching activity 1b (15 minutes)

Identify multiples of numbers 1–12

- Distribute Resource 7: Multiplication grid. Point along the second row: Say: **Two, four, six … What type of numbers are these?** *(even)* **We also call these numbers the multiples of 2.** Refer to the column headed by 2. Ask: **What do you notice? Can you describe the numbers, using the word 'multiple'?** *(They are multiples of 2.)*

- Repeat with 3 and then two further numbers, asking for children's choices if possible, reinforcing the term 'multiple' each time. Discuss the fact that all multiples of 2 are even, all multiples of 5 end in 0 or 5, all multiples of 10 end in 0.

- Show children how the grid works: the answer to 3 × 4 is found where the third row meets the fourth column, and where the fourth row meets the third column, because 3 × 4 = 4 × 3 = 12.

- Ask children to identify multiples in general, and then specific ones. Ask: **What is a two-digit multiple of 7? How many single-digit multiples of 3 are there? What is a multiple of 6 in the 30s? What is the fourth multiple of 5, the eighth multiple of 3?** Watch how the children are working these out: check whether they are counting along the row or down the column, or if they are using the grid in the way you explained earlier.

- Ask: **How could we work out the seventh multiple of 10 without counting?** *(7 × 10)* **The fourth multiple of 9?** *(4 × 9)* **The third multiple of 12?** *(3 × 12)*

- Ask some quickfire questions: **What is the sixth multiple of 3? The tenth multiple of 5?**

Teaching activity 2a (20 minutes)

Identify factors, factor pairs and common factors of two-digit numbers

- Demonstrate how to arrange 12 counters in an array of three rows of four. Record this as 3 × 4 = 12. Tell the children: **12 can be divided into three rows with four counters in each.** Record as 12 ÷ 3 = 4. Turn the array a quarter turn to show four rows of three. Record as 4 × 3 = 12 and 12 ÷ 4 = 3. Explain that numbers that divide into another number exactly, with none left over, are **factors**. So 3 and 4 are factors of 12. Repeat with 2 × 6 and 6 × 2, then with 12 × 1 and 1 × 12.

- Sketch a spider diagram on a whiteboard and record pairs of factors at the ends of a pair of opposite legs. Explain that 3 and 4 are a factor pair of 12 because they multiply together to make 12. Identify the other factor pairs of 12 (1 and 12; 2 and 6).

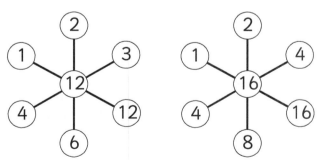

- Give children, in pairs, 20 counters to work out the factors and factor pairs of 20, recording them in a spider diagram, as on the practice page. Ask: **What are the factors of 20? Give me a factor pair of 20. Is 15 a factor of 20?**

- Look at the factor diagrams for 12 and 20 to identify their common factors of 12 and 20: 1, 2 and 4.

Teaching activity 2b (20 minutes)

Identify factors, factor pairs and common factors of two-digit numbers

- Use Resource 7: Multiplication grid to show children how to identify factors of 12. Find 12 in the main body of the grid, go up the column to see which number heads the column. This could be 2, 3, 4, 6 or 12. Say: **2 (or 3, 4, 6, 12) is a factor of 12 because 12 can be divided by 2 exactly with no remainders.** Now trace back along the row to find the other number: (6, 4, 3, 2, 1 respectively). Say: **6 (or 4, 3, 2, 1) is also a factor of 12. 2 × 6 = 12, so 2 and 6 are a** factor pair **of 12 because they multiply together to make 12.** Find 12 in a different place on the grid and identify the factors and the factor pairs. Ask the children to find all the factors pairs (1 × 12, 2 × 6, 3 × 4). Establish that 12 × 1, 6 × 2 and 4 × 3 are all factor pairs of 12. Make a list of all the factors of 12 (1, 2, 3, 4, 6, 12), recording them in a spider diagram.

- Repeat for 20, showing that the factor pairs shown on the grid are 2 × 10 and 4 × 5. Explain that as 1 and the number are always factors of a number, there is another factor pair that does not appear on the multiplication grid: 1 × 20. Make a list of all the factors of 20 (1, 2, 4, 5, 10, 20) and record them in a spider diagram.

- Ask children to identify the numbers that appear in the lists of factors of both 12 **and** of 20 (1, 2 and 4). Explain that these numbers are the **common factors** of 12 and 20 because they are factors of both numbers.

Unit 12: Know and use the vocabulary of prime numbers, prime factors and composite (non-prime) numbers

Content domain reference: 5C5b

Prerequisites for learning

Know and use the language of multiples and factors

Learning outcomes

Identify prime and composite numbers to at least 20

Identify prime factors and express some composite numbers as products of prime numbers

Key vocabulary

Prime number, composite number, prime factor

Resources

Resource 7: Multiplication grid; Resource 8: Prime and composite numbers completed; number cards 2–20; large sheet of paper (A1) with a blank Carroll diagram drawn on it; mini-whiteboards (one per child)

Background knowledge

- **Prime numbers** are the building blocks of all numbers. A prime number has exactly two factors, 1 and itself. 1 is not a prime number as it has only one factor. A prime number cannot be divided exactly by any other numbers than 1 and itself.

- **A composite number** can be divided exactly by numbers other than 1 and itself. Composite numbers have extra factors. Composite means 'made by combining other things'. Every composite number can be expressed as a product of prime numbers.

- A **prime factor** is simply a **factor** that is also a **prime number**.

Teaching activity 1a (10 minutes)

Identify prime and composite numbers to at least 20

- Say to the children: **A prime number is a number that has exactly two factors, 1 and itself. It is not a multiple of any other number except 1 and itself.** Then say: **A composite number is a number that is a multiple of other numbers. It has other factors as well as 1 and itself. It is not a prime number.**

- Show the Carroll diagram, explaining the type of number that goes in each cell.

	prime	composite
even		
odd		

- Take each number card in turn, discuss whether it is a multiple of any other number, counting up in multiples if necessary to decide if it is prime or composite. Ask: **Can this number be divided equally into sets? Does it have any extra factors? Which multiplication table is it in?** Refer back to Unit 11 and Resource 7: Multiplication grid to check, if necessary. Ask: **Where will this number go in our diagram?** If necessary, make a further statement, such as: **It is a prime number and it is odd.** Ask: **Why is 15 (9, 16) not a prime number?** (*It is a multiple of... 3 is a factor...*)

- When all the cards are in place, make a list of the prime numbers under 20.

- Then make a list of the composite numbers to 20.

Teaching activity 1b (10 minutes)

Identify prime and composite numbers to at least 20

- Ask children to write the numbers 1–20 on their mini-whiteboards, in order. Explain that they are going to tick the composite numbers and cross the prime numbers.

- Review the definition of a prime number (a number with exactly two factors, 1 and itself; it is not a multiple of any other number except 1 and itself) and of a composite number (a multiple of other numbers, it has other factors as well as 1 and itself; it is not a prime number).

- Use Resource 7: Multiplication grid, or children's knowledge, to identify and tick the numbers that have more than two factors. Ask: **Is this number a multiple of 2 (or 3, 5, 7)?** If the answer is 'yes', then tick the number. Ask: **What are the factors of 12?** (*1, 2, 3, 4, 6, 12*) **Does it have more than two factors?** (*Yes, so it is a composite number.*) Tick 12.

- Work through all the numbers 1–20, ascertaining that 1 is neither composite nor prime as it only has one factor, that 2, 3, 5, 7, 11, 13, 17 and 19 are prime and 4, 6, 8, 9, 10, 12, 14, 15, 16, 18 and 20 are composite. Write out lists of primes and composites. Ask: **Which is the only even prime number?** (*2*)

Teaching activity 2a (15 minutes)

Identify prime factors and express some composite numbers as products of prime numbers

- Explain that composite numbers can be made by multiplying two or more prime numbers together. Use a systematic approach to demonstrate.

- Ask children to write the **composite numbers** to 20 on their mini-whiteboards.

- Ask children to read out the prime numbers less than 10: 2, 3, 5, 7.

- Multiply 2 by the other four primes (2×2, 2×3, 2×5, 2×7), then ask the children to write the product against the relevant number on their mini-whiteboards ($4 = 2 \times 2$; $6 = 2 \times 3$; $10 = 2 \times 5$; $14 = 2 \times 7$).

- Discuss each composite number. For example, raise the fact that because $6 = 2 \times 3$, 2 and 3 are a factor pair of 6. Both 2 and 3 are prime numbers and factors of 6, so they are called 'prime factors'. Emphasise that 2 and 3 are the prime factors of 6.

  Say: **The prime factors of 10 are 2 and 5. What are the prime factors of 14?** (*2, 7*)

- Multiply 3 by 5 and then by 7. Ask: **What are the prime factors of 15?** (*3, 5*) **Of 21?** (*3, 7*)

- Look together at the composite numbers remaining (8, 12, 16, 18 and 20). Explain that these can be found by multiplying more than two prime numbers together.

- Work through some examples with three prime numbers, for example: $2 \times 2 \times 2 = 8$, $3 \times 3 \times 3 = 27$; then $2 \times 2 \times 3 = 12$, $2 \times 2 \times 5 = 2$, $2 \times 3 \times 3 = 18$. Then give some examples of multiplying four numbers: $2 \times 2 \times 2 \times 2 = 16$.

- Children could use Resource 8: Prime and composite numbers completed to explore numbers greater than 20. Ask: **What is 18 as a product of prime numbers?** ($2 \times 3 \times 3$) **Which is the only even prime number?** (*2*)

Teaching activity 2b (15 minutes)

Identify prime factors and express some composite numbers as products of prime numbers

- Children, in pairs, draw spider diagrams to show the factor pairs of each **composite** number to 20, dividing the work between them. (See Unit 11.)

- Ask them to underline or circle the factors that are prime numbers. Explain that factors that are also prime numbers are called prime factors.

- Say that all the composite numbers can be made by multiplying two or more prime numbers together. Demonstrate, using the factors of 6: 2 and 3 are prime factors of 6 and $2 \times 3 = 6$. Now use the factors of 12. Demonstrate that 2 and 3 are prime factors of 12: $2 \times 3 = 6$. Ask: **What other number do we need to multiply by, to make 12?** (*2: we can write $12 = 2 \times 2 \times 3$.*) Say: **This is called writing a number as a product of prime factors.**

- Ask the children to explore which numbers can be made just by multiplying their prime factors together. Can the children make the other numbers just from combinations of their prime factors? They could explore numbers greater than 20.

Unit 13: Establish whether a number up to 100 is prime and recall prime numbers up to 19

Content domain reference: 5C5c

Prerequisites for learning

Know and use the language of multiples and factors

Identify prime numbers to 20

Learning outcomes

Identify prime and composite numbers to 100

Key vocabulary

Prime number, composite number, prime factor

Resources

Resource 6: 100 square; Resource 7: Multiplication grid; Resource 8: Prime and composite numbers completed

Background knowledge

Prime numbers are the building blocks of all numbers. A prime number has exactly two factors, 1 and itself. 1 is not a prime number as it has only one factor. A prime number cannot be divided exactly by any other number other than 1 and itself.

The Sieve of Eratosthenes is an old method for finding all the prime numbers to 100. It 'drains out' composite numbers and leaves prime numbers behind.

There are 25 prime numbers less than 100: 2, 3, 5, 7, 11, 13, 17, 19, 23, 29, 31, 37, 41, 43, 47, 53, 59, 61, 67, 71, 73, 79, 83, 89, 97.

Teaching activity 1a (20 minutes)

Identify prime and composite numbers to 100

- Ask children to remind you what prime numbers and composite numbers are. Confirm that every number (apart from 1) has at least two factors, 1 and itself, but most numbers have **extra factors**. If a number has extra factors, then it is not a prime number, it is a **composite** number. **Why is 1 not a prime number? How do we know that 15 is not a prime number?** (*It has more factors than 1 and itself.*) **Which numbers to 20 are prime?** (*2, 3, 5, 7, 11, 13, 17, 19*)

- Give each child a copy of Resource 6: 100 square and explain that they are going to use a method used by the Ancient Greeks to find all the prime numbers to 100. Say: **It is called the Sieve of Eratosthenes, as he was the Greek who invented it. We are going to use our knowledge of multiples to help us cross out all the composite numbers, leaving just the prime numbers in the sieve on the 100 square.**

- Tell the children to start by crossing out 1, as it is not prime or composite. Tell them to circle 2, then cross out all the multiples of 2, starting with 4. They should see that this eliminates alternate columns – all the even numbers – as they all have a factor of 2 at least.

- Next, circle 3 and then cross out the multiples of 3, continuing to 99 by counting on in 3s. Every alternate multiple of 3 (the even multiples) will already have been crossed out. Ask: **What about 4? Are there any multiples of 4 left to cross off?** (*No as all multiples of 4 are also multiples of 2.*) Continue with 5, crossing out all its multiples, which are all in two columns, but note that the 10s column has already gone. Continue with multiples of 7. There should be 25 circled numbers left: 2, 3, 5, 7, 11, 13, 17, 19, 23, 29, 31, 37, 41, 43, 47, 53, 59, 61, 67, 71, 73, 79, 83, 89, 97. These are all primes as they are not in any multiplication tables except their own.

Teaching activity 1b (20 minutes)

Identify prime and composite numbers to 100

- Distribute copies of Resource 6: 100 square and Resource 7: Multiplication grid. Say: **You are going to identify all the prime numbers between 1 and 100 by crossing out all the composite numbers on a 100 square.** Ask: **Which numbers to 20 are prime?** (*2, 3, 5, 7, 11, 13, 17, 19*) **Which are composite?** (*all the others except 1*)

- Tell children to cross through the first row and column on the multiplication grid, as 1 is not a prime number. They are now going to cross out the numbers on the 100 square that **are** in the multiplication table. They must start at the **2×** row. This work could be divided up to save time.

- Discuss the fact that some numbers have already been crossed off as they are multiples of both 2 and 3, 3 and 5, ….

- There will still be some composite numbers left, as they are beyond the twelfth multiple. Discuss which ones can be crossed off easily. (*all remaining even numbers and multiples of 5 ending in 5*)

- Explain that we can check multiples of 3 above 36 by adding the digits. If the digits of a number add to a multiple of 3, then that number is also a multiple of 3. Use this rule to check remaining numbers larger than 36. Alternatively, count on in 3s from 36. This will eliminate 39, 51, 57, 87 and 93.

- Finally, check if any of the remaining numbers are divisible by 7 by counting on from 84 (12 × 7).

- There should be 25 numbers left: 2, 3, 5, 7, 11, 13, 17, 19, 23, 29, 31, 37, 41, 43, 47, 53, 59, 61, 67, 71, 73, 79, 83, 89, 97. These are all primes as they are not in any multiplication tables except their own. Children should circle all the prime numbers and check that they have 25.

Unit 14: Multiply and divide numbers mentally, drawing upon known facts

Content domain reference: 5C6a

Prerequisites for learning

Recall multiplication and division facts for multiplication tables to 12 × 12

Multiply and divide by 10 and 100, giving whole-number answers

Learning outcomes

Multiply numbers mentally, drawing upon known facts

Divide numbers mentally, drawing upon known facts

Key vocabulary

Related fact, bar modelling, inverse, divisor

Resources

Resource 9: Bar models; board pens in at least two different colours; coloured pencils

Background knowledge

Although there is no longer a specific mental maths test in the KS2 SATs, children still need to be able to work out calculations mentally because this is often much quicker than using a formal written method. It can be used to check that their answer makes sense. Often, children do not immediately see the connection between, for example, 3 × 6 and 30 × 60.

Teaching activity 1a (15 minutes)

Multiply numbers mentally, drawing upon known facts

- Begin by writing a two-digit number on a whiteboard and asking children to double it. If they struggle, show them how to partition the number.

  37 = 30 + 7… double it … 60 + 14 = 74

  76 = 70 + 6 … double it … 140 + 12 = 152

- Ask: **What number are we multiplying by when we double a number?** (2)

- Write a multiplication fact such as 4 × 6 = 24 in the middle of a whiteboard. Ask: **Which other multiplication facts can we work out from this one?** Use a different coloured pen for the related fact: consider **4 × 60 = 4 × 6 × 10 = 24 × 10 = 240** and **40 × 60 = 4 × 6 × 10 × 10**. Ask: **Can you find six related facts for 4 × 6 = 24?**

- Write the basic fact in one colour and the zeros in another. Let the children write the facts down on paper, using a coloured pencil for the known fact. Tell them to find six facts for each multiplication fact they choose.

- If children are still struggling, use Resource 9: Bar models to illustrate single digit × multiple of 10, 100, 1000, for example:

  4×60 | 60 | 60 | 60 | 60 | = 240

Teaching activity 1b (15 minutes)

Multiply numbers mentally, drawing upon known facts

- Write a multiplication fact such as 4 × 6 = 24 in the middle of a whiteboard. Circle it and draw six legs to make a spider diagram. At the ends of the legs write these calculations: 4 × 60 = ; 40 × 6 = ; 40 × 60 = ; 0·4 × 6 = ; 4 × 600 = ; 4000 × 6 = . Say: **Because we know 4 × 6 = 24, we can work out these harder calculations mentally.** Start by writing 24 after each equals sign. Say: **We know that the digit 2 followed by 4 will be in the answer somewhere.** Point to 40 × 60 = . Ask: **How many zeros are there in this question?** (2) **So we need the same number of zeros in our answer.** Complete the calculation: **40 × 60 = 2400.** Use two colours to differentiate the facts from the zeros if that is helpful.

- Repeat with a fact that ends in zero, such as 5 × 6 = 30. Then stress, for example, that 50 × 60 needs two **extra** zeros, giving 3000. Say: **This is because 5 × 6 = 30.** Using two colours to differentiate between the 30 and the extra two zeros will highlight this.

Teaching activity 2a (15 minutes)

Divide numbers mentally, drawing upon known facts

- Write a two-digit number on a whiteboard and ask children to halve it. If they struggle, show them how to partition the number.

  $76 = 70 + 6$

  half of 70 is 35

  half of $6 = 3$

  $35 + 3 = 38$

- Use Resource 9: Bar models to illustrate some division facts, such as $28 \div 7$. Count up in 7s to 28. Four 7s make 28: $4 \times 7 = 28$ so $28 \div 7 = 4$.

28			
7	7	7	7
7	14	21	28

- Ask: **Which other related division fact do we know?** ($28 \div 4 = 7$)

- Repeat for related facts $280 \div 7$; $2800 \div 4$, referring the children back to the multiplications in Activity 1a or 1b and noting the extra zeros in the question.

- Now consider $280 \div 70$. Count up in 70s. Ask: **How many 70s make 280?** Establish that it is only four because the zero in the number being divided is cancelled out by the zero in the dividing number (the divisor).

- Repeat for several other division facts, writing the **two** related divisions and continuing with divisions of multiples of 10 and 100.

- Repeat but use a missing number division: $32 \div \square = 8$. Ask: **How can we rewrite this?** ($32 \div 8 = \square$ or $\square \times 8 = 32$, so $\square = 4$) Repeat with more examples, including some multiples of 10 and 100, such as $1800 \div \square = 9$.

Teaching activity 2b (15 minutes)

Divide numbers mentally drawing upon known facts

- Explain that division is the inverse of multiplication, which means that division 'undoes' the multiplication. Write a multiplication fact such as $3 \times 5 = 15$. Ask: **Which related division facts do we know?** ($15 \div 3 = 5$) **Which division facts will undo this multiplication?** ($15 \div 5 = 3$ and $15 \div 3 = 5$) Repeat with other multiplication facts if necessary.

- Give children some division tables facts to answer mentally.

- Repeat with a multiple of 10 and 100 multiplication fact, such as $500 \times 3 = 1500$, ascertaining that $1500 \div 3 = 500$ and $1500 \div 500 = 3$.

- Write a division fact such as $15 \div 5 = 3$ in the middle of a whiteboard, circle it and draw six legs to make a spider diagram. Ask: **What is the other related division fact we also know?** ($15 \div 3 = 5$) Write this in the centre too. At the ends of the legs write these calculations: $150 \div 5$; $150 \div 30$; $1500 \div 5$; $1500 \div 50$; $1500 \div 30$; $15000 \div 500$.

- Highlight the numbers of the original division fact: $15 \div 5$ or $15 \div 3$. Explain that we can use our basic division fact to answer these related divisions easily. Ask: **Will the answer to $150 \div 5$ be larger or smaller than $15 \div 5$?** (larger) When children have an answer, they should check with a related multiplication that it makes sense: $150 \div 5 = 30$, $5 \times 30 = 150$ ✓

- Spend time on facts with zeros in both numbers: in division, these cancel out.

- Record the division with the last zeros crossed out, for example: $150\cancel{0} \div 3\cancel{0} = 150 \div 3 = 50$. Check with related multiplication that the answer is sensible: $1500 \div 30 = 50$; $30 \times 50 = 1500$.

Unit 15: Multiply and divide whole numbers and those involving decimals by 10, 100 and 1000

Content domain reference: 5C6b

Prerequisites for learning

Multiply whole numbers by 10

Divide multiples of 10 and 100 by 10

Understand the concept of tenths and hundredths as decimals

Know that ten tenths = 1 whole one and ten hundredths = 1 tenth

Learning outcomes

Multiply any number by 10, 100 and 1000

Divide any number by 10, 100 and 1000

Key vocabulary

Place value, decimal point, tenths, hundredths

Resources

Resource 10: Decimal place-value grids 1 (one per child and two enlarged versions); Resource 11: Decimal place-value grids 2; digit cards 0–9 (one set per pair or per child)

Background knowledge

When you multiply or divide any number by a power of 10 (10, 100, 1000 …) the value of each digit changes, becoming 10 (or 100 or 1000) times larger or smaller. This means that the place value of each digit moves up or down by the same number of places as there are zeros in the power of ten.

It is important for children to understand that the order of the digits stays the same. Zeros can only be used as place holders that will always be at the end of the number in whole numbers, and just before or after the decimal point in numbers involving decimals; for example: 234 × 100 = 23 400; 23 ÷ 1000 = 0·023.

Children often think that they can just 'add a zero', but since this doesn't work with decimals, it is better to talk about zeros as place holders in whole numbers. Although the decimal point does not ever move (it stays between the ones and tenths) children often find it easier to 'move it' the relevant number of places.

Teaching activity 1a (15 minutes)

Multiply any number by 10, 100 and 1000

- Using Resource 10: Decimal place-value grids 1, explain that the children will be using a place-value grid to see what happens when numbers are multiplied by 10, 100 and 1000. Use the enlarged grid and ask a child to suggest a three-digit whole number, such as 265, and to write it in the correct place in the grid. Write ×10 in the far right-hand column of the next row. Ask: **What is (265) × 10? Who can write this in the correct place?** Point out that the value of each digit has become ten times larger. Say: **The 5 has become 50, the 60 is now 600 and the 200 is now 2000.** Now write ×100 in the next row and repeat, ascertaining that each digit is worth 100 times more. Ask: **What is the value of the 6 now?** Repeat for ×1000.

- Establish that each digit moves the same number of places left as the number of zeros in the multiplier.

- Repeat with a two-digit number with one decimal place, such as 34·5, stressing that 10 × tenths takes the digit to the next column to the left, the ones. Ask: **What is the value of 5 in the original number? What is the value of the 5 now?** Stress that we cannot just 'add a zero'. Write a zero in the hundredths place and discuss what has happened to the number. Say: **It has not changed; each digit still has the same value. Each digit must move one place.**

- Repeat with a variety of numbers with two decimal places, such as 376·19, 14·72, 3·58.

Teaching activity 1b (15 minutes)

Multiply any number by 10, 100 and 1000

- Use the digit cards to make a three-digit number such as 135. Ask a child to put the digits in the correct place on Resource 11: Decimal place-value grids 2. Ask: **Where is the decimal point?** (*At the end of a whole number, although we do not usually write it unless there are decimal places.*) Start with ×10. Ask: **What happens to each digit when we multiply this number by 10?** (*They all move one place to the left.*) Move each digit one place to the left. Ask: **There is a space here; what do we need?** (*A zero as a place holder because when we multiply a three-digit number by ten, the result is a four-digit number.*) Put a zero as a place holder and read the new number together. Ask; **What is the value of the (3) now?** (*300*) Repeat by multiplying the original number by 100 and then by 1000 and moving each digit two places or three places to the left and adding two zeros or three zeros as place holders. Record each calculation, for example: 234 × 10 = 2340. Say: **The order of the digits does not change; it is as if they are glued together.**

- Repeat with a two-digit number with one decimal place, such as 45·6. Stress that we cannot just 'add a zero'. Put a zero in the hundredths column and discuss what has happened to the number. Say: **It has not changed; each digit is still the same value. Each digit must move one place.** Point out that ×10 does not need a zero place holder this time, as the 0·5 jumps one place over the decimal point to become 5. With ×100 the 0·5 becomes 50, so a zero is needed as a place holder in the ones column.

- Repeat with a variety of numbers with two decimal places, such as 376·19, 14·72, 3·58, discussing the new value of each digit.

Teaching activity 2a (15 minutes)

Divide any number by 10, 100 and 1000

- Display Resource 10: Decimal place value grids 1 and explain that the children will be using a place-value grid to see what happens when numbers are divided by 10, 100 and 1000. Establish that the value of each digit becomes 10, 100, 1000 times smaller. Divide these numbers by 10. Start with a four-digit number (7523), then a three-digit number (523), then a two-digit number (23). Ask: **What is the value of the 3 now? The 3 was in the ones column before we divided by 10. Which column is it in now?** (*tenths*) Show children how to use zeros as place holders at the beginning of numbers where necessary: 23 ÷ 100 = **0·**23, 23 ÷ 1000 = **0·0**23. You may need to explain that the 3 is thousandths, the third decimal place.

- Repeat with a variety of four, three- and two-digit numbers with one decimal place, just dividing by 10 and 100.

Teaching activity 2b (15 minutes)

Divide any number by 10, 100 and 1000

- Use Resource 11: Decimal place value grids 2 and digit cards to demonstrate dividing by 10, 100, 1000. Establish that the value of each digit becomes 10, 100, 1000 times smaller and that each digit must move to the right. Start with a four-digit number (8542). Ask a child to put the digits in the correct place on the resource sheet. Ask: **Where is the decimal point?** (*At the end of the whole number, although we do not usually write it unless there are decimal places.*) Start with ÷ 10. Ask: **What happens to each digit when we divide this number by 10?** (*They all move one place to the right.*) Move each digit one place to the right. Ask: **What is the value of the 4 now? The 2 was in the ones column before we divided by 10. Which column is it in now?** Repeat by dividing the same number by 100 and 1000.

- Repeat with three-digit numbers, then two-digit numbers, showing how to use zeros as place holders at the beginning of numbers where necessary: 45 ÷ 100 = **0·**45, 28 ÷ 1000 = **0·0**28. You may need to explain that the 8 is thousandths, the third decimal place.

- Repeat with a variety of two-, three- and four-digit numbers with one decimal place, just dividing by 10 and 100.

Unit 16: Multiply numbers up to four digits by a one- or two-digit number using a formal written method, including long multiplication for two-digit numbers

Content domain reference: 5C7a

Prerequisites for learning

Recall multiplication facts for multiplication tables to 12 × 12

Work out mentally multiples of 10 and 100

Multiply and divide by 10

Multiply two- and three-digit numbers by a one-digit number

Key vocabulary

Column method, formal method, multiplication, product

Resources

Resource 7: Multiplication grid (optional)

Learning outcomes

Use formal column methods for short multiplication

Use formal column methods for long multiplication

Background knowledge

The new curriculum specifies traditional, formal column methods for multiplication, so a transition is needed from the popular grid method or partitioning method to a column method. An interim step is the expanded column method shown in activities 1b and 2b. In order to learn the methods, some children may need the support of a multiplication grid. Alternatively, use multipliers of 2, 3, 4 and 5 to teach the methods. In long multiplication, children need to understand that each digit in the top row must be multiplied by each digit in the second row.

Teaching activity 1a (20 minutes)

Use formal column methods for short multiplication

- Begin with a few quickfire questions on multiplication facts, using a variety of language, for example: 3 times 7, 5 multiplied by 8, the product of 6 and 7. Test the less secure 6, 7, 8 and 9 times tables.

- Remind children that multiplication can be done in any order. Say: **If you know 8 × 5 then you know 5 × 8.**

- Write 375 × 4 on a whiteboard. If children do not already use a column method, explain that they now need to use what is a called a 'formal written method', in which they work out the multiplication in columns. Explain that, for this method, they only need to know multiplication facts to 9 × 9 because in this method you only multiply a single digit by a single digit. Explain that when they write the numbers in columns, the digits appear in the correct place-value columns.

- Work through the short multiplication method.

$$
\begin{array}{r}
\ 3\ \ 7\ \ 5 \\
\times\ 4 \\
\hline
1\ \ 5\ \ 0\ \ 0 \\
3\ \ 2 \\
\hline
\end{array}
$$

$4 \times 5 = 20$ write the 0 in the units answer column, carry the 2 tens.

$4 \times 7 = 28$, plus the carried 2 gives 30, write zero in the answer, carry 3 over.

$4 \times 3 = 12$, plus the carried 3 gives 15, write 15 in the answer line.

Repeat with a four-digit number multiplied by a one-digit number: 1639 × 8.

$$
\begin{array}{r}
1\ \ 6\ \ 3\ \ 9 \\
\times\ 8 \\
\hline
1\ \ 3\ \ 1\ \ 1\ \ 2 \\
5\ \ 3\ \ 7 \\
\hline
\end{array}
$$

$8 \times 9 = 72$; $8 \times 3 + 7 = 24 + 7 = 31$;

$8 \times 6 + 3 = 48 + 3 = 51$

$8 \times 1 + 5 = 13$

Teaching activity 1b (20 minutes)

Use formal column methods for short multiplication

- Begin with a few quickfire questions practising multiples of 10 and 100 multiplied by numbers 1 to 9, such as: 40 × 5; 9 × 60 (Unit 15). Follow on with multiples of 10 multiplied by multiples of 10, such as: 30 × 50; then 300 × 20.

- Write 258 × 4 on the board. If the children do not already use a column method, explain that they now need to use what is a called a 'formal written method' that is worked out in columns, but that it is very similar to both the grid and partitioning methods. This is the expanded column method: 3-digit × 1-digit needs 3 × 1 parts; 4 digit × 1 digit needs 4 × 1 parts. Work through these examples, letting children refer to Resource 7: Multiplication grid if necessary:

```
      2 5 8
  ×       4
      3 2     4 × 8 = 32
    2 0 0     4 × 50 = 200
    8 0 0     4 × 200 = 800
  1 0 3 2     Add them together
    2
```

```
    2 7 5 6
  ×       9
        5 4     9 × 6
      4 5 0     9 × 50
    6 3 0 0     9 × 700
  1 8 0 0 0     9 × 2000
  2 4 8 0 4
  1     1
```

Teaching activity 2a (20 minutes)

Use formal column methods for long multiplication

- Explain that the same column method can be used when multiplying a number with two or more digits by a two-digit number, but we need an extra answer line and a 'zero place holder' for multiplying by a tens number. Work through the following examples, paying particular attention to the second line when multiplying by the 10s. Start with 2563 × 18, first multiplying by 8. Then work out the second line, which is simply 2563 × 10, showing how to use the zero place holder. Next, work through 7296 × 24. For the second line, stress that 20 is 10 × 2, so write the zero place holder in the ones column and continue as with short multiplication, 2 × 7296. Allow plenty of space for the carried numbers, explaining that these numbers should be written smaller than the other numbers so that it is clear which digits have to be added and which are the answers.

Alternatively, the small carried numbers could be written above. Check the method your school uses or use whatever is better for the children.

```
    2 5 6 3
  ×     1 8
  2 0 5 0 4     8 × 2563 as
    4 5 2        short method
  2 5 6 3 0     10 × 2563
  4 6 1 3 4     add together
  1
```

```
      7 2 9 6
  ×       2 4
    2 9 1 8 4     4 × 7296
      1 3 2
  1 4 5 9 2 0     10 × 2 × 7296
      1 1
  1 7 5 1 0 4     add together
  1 1 1
```

Teaching activity 2a (20 minutes)

Use formal column methods for long multiplication

- Explain that the same expanded column method can be used when multiplying by a 2-digit number, but extra answer lines are needed, together with special care when multiplying by the tens number: 4-digit × 2-digit needs 4 × 2 calculations. Work through some examples:

```
      2 5 6 3
  ×       1 8
        2 4     8 × 3
      4 8 0     8 × 60
    4 0 0 0     8 × 500
  1 6 0 0 0     8 × 2000
        3 0     10 × 3
      6 0 0     10 × 60
    5 0 0 0     10 × 500
  2 0 0 0 0     10 × 2000
  4 6 1 3 4
  1 1 1
```

```
      7 2 9 6
  ×       2 4
        2 4       4 × 6
      3 6 0       4 × 90
      8 0 0       4 × 200
  2 8 0 0 0       4 × 7000
      1 2 0       20 × 6
    1 8 0 0       20 × 90
    4 0 0 0       20 × 200
1 4 0 0 0 0       20 × 7000
1 7 5 1 0 4
  1 2 1
```

Unit 17: Divide numbers up to four digits by a one-digit number, using the formal written method of short division, and interpret remainders appropriately for the context

Domain content reference: 5C7b

Prerequisites for learning

Recall division facts for multiplication tables to 12 × 12

Multiply whole numbers by 10 and 100

Learning outcomes

Use a formal written method for short division

Key vocabulary

Divide, divisor, remainder, repeated subtraction

Resources

Whiteboard or a large sheet of paper

Background knowledge

The new curriculum favours the traditional formal method of division, as it is very efficient. It is sometimes called the **bus-stop method**. However, it can be challenging for children to understand. The most common areas of confusion are:

• where to put the answer and where to put the remainder

• using a lower multiple, resulting in the next number being too high

• omitting the remainder when the divisor does not divide into the first digit, or omitting zeros.

Look out for examples of these errors, as illustrated in the second example in activity 1a.

Repeated subtraction (activity 1b) is an alternative acceptable method, but it takes more time. Encourage children who struggle to work with the bus-stop method to use this method if they are more successful with it.

Teaching activity 1a (20 minutes)

Use a formal written method for short division

• As a warm-up, practise some division facts with the children, such as: 32 divided by 8.

• Show the children how to set up a division, such as 4185 ÷ 5, using the traditional 'bus-stop method' and the commentary suggested.

$$\begin{array}{r} 8 \quad 3 \quad 7 \\ 5 \overline{\smash{)}\, 4 \;\; {}^{4}1 \;\; {}^{1}8 \;\; {}^{3}5} \end{array}$$

• Give a running commentary as you work through the division: **4 is not divisible by 5, so carry the 4 over to the next column, writing it as a small number before the 1; now we have 41. What is 41 ÷ 5?** (*8 remainder 1*) **Write the 8 on the answer line above the hundreds column. Write the remainder in the next column, as a small number before the 8, giving 18. What is 18 ÷ 5?** (*3 remainder 3*) **Write the first 3 on the answer line above the tens column and write the remainder 3 in the**

next column, as a small number before the 5, giving 35. What is 35 ÷ 5? (*exactly 7 with no remainder*) **Write 7 on the answer line in the ones column.**

• Repeat with more examples with no remainder: 2564 ÷ 4; 2175 ÷ 3; 4284 ÷ 6.

• Now work through an example that will have a remainder, such as 7625 ÷ 4. Ask: **How can we tell that this division will have a remainder?** (*It is an odd number divided by an even number.*)

$$\begin{array}{r} 1 \quad 9 \quad 0 \quad 6 \quad \text{remainder } 1 \\ 4 \overline{\smash{)}\, 7 \;\; {}^{3}6 \;\; 2 \;\; {}^{2}5} \end{array}$$

• Work through the calculation, finally showing the remainder as a whole number. Give a running commentary as you work through the division: **7 ÷ 4 = 1 remainder 3; 36 ÷ 4 = 9, no remainder; 2 is not divisible by 4, so write 0 on the answer line above the tens column. 25 ÷ 4 = 6, remainder 1.**

- Repeat with more examples with a remainder, discussing how we know there will be a remainder: 2856 ÷ 5 (*2856 is not a multiple of 5.*); 2847 ÷ 8 (*2847 is an odd number; multiples of 8 are even.*); 2857 ÷ 9 (*The digits do not add to a multiple of 9.*)

Teaching activity 1b (20 minutes)

Use a formal written method for short division

- Practise some division facts to warm up, such as 32 divided by 8.

- Say: **You are going to work out some more challenging divisions, using repeated subtraction of multiples of the divisor.**

- Write 865 ÷ 5 at the top of a whiteboard or large sheet of paper. Explain that repeatedly subtracting just 5 would take a long time, so instead we can subtract larger multiples of 5 that we know. Show children how to write some multiples based on the coins they know (1p, 2p, 5p, 10p ...): $1 \times 5 = 5$, $2 \times 5 = 10$, $5 \times 5 = 25$, $10 \times 5 = 50$, $20 \times 5 = 100$, $50 \times 5 = 250$, $100 \times 5 = 500$, $200 \times 5 = 1000$. Ask the children to pick out the largest of these multiples that can be subtracted each time, recording each step.

```
    8  6  5
 -  5  0  0     100 × (200 × is
 ─────────          more than 865)
    3  6  5
 -  2  5  0     50 ×
 ─────────
    1  1  5
 -  1  0  0     20 ×
 ─────────
       1  5
 -     1  0     2 × (if you know 3 × 5 = 15,
 ─────────          you could do this in one step,
          5         but we do not have a 3p coin!)
 -        5     1 ×
 ─────────
          0     173
```

- Repeat with more examples without remainders, extending to four-digit numbers, such as: 2564 ÷ 4; 2175 ÷ 3; 4284 ÷ 6. If necessary, you can use 500 and 1000 times the divisor to subtract.

- Now repeat, showing an example with a remainder. Write out the 'coin' multiplications first: 7625 ÷ 4: 4 (1), 8 (2), 20 (5), 40 (10), 80 (20), 200 (50), 400 (100), 2000 (500), 4000 (1000).

- Explain that children can also use multiples of these numbers and their knowledge of the 4× table to cut down some of the steps.

```
    7  6  2  5
 -  4  0  0  0    1000
 ──────────────
    3  6  2  5
 -  2  0  0  0    500
 ──────────────
    1  6  2  5
 -  1  6  0  0    400 (4 × 100) some
 ──────────────      children may need to
          2  5       subtract 400 four
 -        2  4       times instead
 ──────────────         6
             1    1906  2564 ÷ 4 =
                            1906 remainder 1
```

- Repeat with more examples **with** remainders, such as: 2856 ÷ 5; 2847 ÷ 8; 2857 ÷ 9, writing out the relevant multiples each time and encouraging children to spot multiples of multiples to cut down the number of subtractions.

Unit 18: Recognise and use square numbers and cube numbers, and the notation for squared (^{2}) and cubed (^{3})

Content domain reference: 5C5d

Prerequisites for learning

Recall of multiplication facts to 12 × 12
Multiply three numbers together

Key vocabulary

Square number, cube number, notation, index, power, base number

Learning outcomes

Recognise and use square numbers and use the correct notation x^2

Recognise and use cube numbers and use the correct notation x^3

Resources

Squared paper; small square tiles or squares of paper; small building cubes; coloured pencils (optional); digit cards 0–9

Background knowledge

Children will have met the first 12 square numbers from multiplication tables. They appear in the leading diagonal, top left to bottom right, in the multiplication grid (see Resource 7: Multiplication grid).

Every number, including decimals and negative numbers, can be squared and cubed.

A square number is formed by multiplying a number by itself (10 × 10). A cube number is formed when one number is used in a multiplication three times (10 × 10 × 10).

Only whole numbers give **perfect** square and cube numbers. The activities make a clear link with area and volume.

Teaching activity 1a (15 minutes)

Recognise and use square numbers and use the correct notation x^2

- Explain that children are going to work with some special numbers, called **square numbers**. They will draw diagrams to show these special numbers.

- Distribute squared paper and ask children to colour just one square at the top left of their paper. Ask them to write the numeral 1 under it. Now ask them to draw a 2 by 2 square next to it. Ask: **How many small squares have you coloured?** (4) Say: **Write 4 underneath your diagram.** Repeat with 3 by 3, 4 by 4 and 5 by 5 squares. Tell children to leave several squares under each diagram for later use.

- Explain that the numbers they have written (1, 4, 9, 16 and 25) are the first five square numbers. Direct children to look carefully at the numbers of rows and columns in each diagram. Elicit that the second square number has two rows and two columns, and 2 × 2 = 4. Ask: **What can you say about the other square numbers you have drawn?** (*third square number: three rows, three columns, 3 × 3 = 9...*) Ask: **How could we work out the sixth square number without drawing it?** (*Follow the pattern: six rows, six columns, 6 × 6 = 36.*) **The tenth square number?** (*Follow the pattern: 10 rows, 10 columns, 10 × 10 = 100.*) Do not write the others down yet, but tell children to write 1 × 1, 2 × 2, 3 × 3... under their diagram.

- Now explain that mathematicians have a special way of writing squared numbers, with a small raised digit: 1^2, 2^2, 3^2, 4^2, 5^2. Let the children write 1^2, 2^2, 3^2, 4^2, 5^2 under the relevant diagrams.

- Encourage children to write down the square numbers to 12^2 without using diagrams unless they need to.

  144

  12×12

  12^2

- Finally, write out the first 12 square numbers in full, for example: $1^2 = 1 \times 1 = 1$, $2^2 = 2 \times 2 = 4$…

Teaching activity 1b (20 minutes)

Recognise and use square numbers and use the correct notation x^2

- Explain that they are going to work with some special numbers, called **square numbers,** and they are going to use squares or tiles to show these special numbers.

- Place one tile on the table and explain that this shows the first square number: one. Put a digit card 1 next to the tile. Ask a child to arrange four tiles into a square: two rows of two tiles. Say: **Four is the second square number.** Put a digit card 4 next to these tiles. Ask: **How many rows will we need for the third square?** (3) **How many tiles will there be in each row?** (3) Ask a child to build the 3 by 3 square. Ask: **How many tiles are there?** (9) Say: **Nine is the third square number.** Repeat for a 4 by 4 square and a 5 by 5 square, using digit cards to place the appropriate number with each array. Explain that these tiles show the first five square numbers: 1, 4, 9, 16, 25.

- Ask: **How can we use multiplication facts to work out the next square number?** (*Six rows of six is 36 tiles. 36 is the next square number.*) Encourage children to work out the square numbers up to 10×10 without using tiles, unless they need to.

- Explain the notation and write out the first 10 square numbers as: $1^2 = 1 \times 1 = 1$, $2^2 = 2 \times 2 = 4$…

Teaching activity 2a (20 minutes)

Recognise and use cube numbers and use the correct notation x^3

- Explain that there are some more special numbers called **cube numbers** and that they are linked to cubes, which have three dimensions. Build a 2 by 2 by 2 cube for children to see. Say: **Two rows of two, which is four in the bottom layer, then another layer. We have two layers, each with four cubes: $2 \times 4 = 8$.** Point to the width, the depth then the height of the cubes as you say: **Two times two times two, so eight altogether. This is the second cube number.** Write: $2 \times 2 \times 2 = 8$.

- Explain that mathematicians also have a special way of writing cube numbers. Ask: **Can anyone guess what it is?** Write $2^3 = 2 \times 2 \times 2 = 8$. Explain that the larger digit, the **base**, is the one being multiplied (2) and the smaller digit, the **power**, is how many times it appears in the multiplication (3).

- Work out the first five cube numbers (three of a number multiplied together). recording as: $1^3 = 1 \times 1 \times 1 = 1$, $2^3 = 2 \times 2 \times 2 = 8$ …

- Ask: **How can we work out the tenth cube number?** (*$10 \times 10 \times 10 = 1000 = 10^3$*)

Teaching activity 2b (20 minutes)

Recognise and use cube numbers and use the correct notation x^3

- Explain that there are some more special number called **cube numbers** and that they are linked to cubes, which have three dimensions. Allow the children to build their own cubes to 5 by 5 by 5. Encourage them to work out the number of cubes in the first layer of each cube, either by counting or multiplication, then multiply this by the number of layers (the height) or by counting up in multiples of the first layer rather than counting each cube. Use digit cards to show the number of small cubes in each large cube.

- When they have worked out the first five cube numbers, record:
  $1^3 = 1 \times 1 \times 1 = 1$, $2^3 = 2 \times 2 \times 2 = 8$ …

- Establish that a cube number is the result of multiplying three of the same number together. Ask: **How can we work out the tenth cube number without building a cube?** (*$10 \times 10 \times 10 = 1000$*) **How would mathematicians record the tenth cube number?** (*10^3*)

Unit 19: Solve problems involving multiplication and division, including using their knowledge of factors and multiples, squares and cubes

Content domain reference: 5C8a

Prerequisites for learning

Multiply numbers with up to four digits by one- and two-digit numbers

Divide numbers with up to four digits by one-digit numbers

Know and use the language of factors, multiples, square and cube numbers

Learning outcomes

Solve problems involving multiplication, division, factors, multiples, square and cube numbers

Key vocabulary

Population, square, cube, multiple, factor

Resources

Resource 12: Multiplication and division word problems, cut into cards

Background knowledge

Before they start this unit and attempt word problems, children need a secure understanding of the previous units in this domain. To solve word problems, children must interpret them and be able to decide which operations they need to use.

Children should be encouraged to write down every calculation they do, even if they do it mentally, in preparation for formal tests where 'method marks' are often awarded.

Teaching activity 1a (30 minutes)

Solve problems involving multiplication, division, factors, multiples, square and cube numbers

- Divide the cards cut from Resource 12: Multiplication and division word problems into separate piles: one of questions and one of operations.
- Work with the children to try to match the operations to the questions, discussing the words that give the clues. As there is sometimes more than one way of solving a problem there may be some overlap, but they should try to match one operation card to each problem. Help them to match those that are more obvious, then see how the others could fit. (The best fit is 1D, 2H, 3E, 4B, 5C, 6F, 7G, 8A.)
- Work through each problem, recording **every** calculation, even if children can work it out mentally, explaining that this is good practice for word problems with more complex calculations. Here are some key questions.

1 **If a teacher wanted pairs for a gym lesson, could she divide the class of 30 exactly into pairs?** (*Yes.*) **How many pairs?** (*15 × 2 = 15 pairs*) **Could she have groups of four without any being left out?**

(*No, 4 is not a factor of 30: 1 × 30; 2 × 15; 3 × 10; 5 × 6; 6 × 5; 10 × 3; 15 × 2; 30 × 1.*)

2 **What do we need to work out first?** (*How many children can go: 11 × 4 = 44.*) **What next?** (*How many teams of five can be made from 44 children: 44 ÷ 5 = 8 remainder 4.*)

3 **We have decided this question is about multiples; multiples of what?** (*7*) **Read the question carefully; which multiples of 7 do we need?** (*Those between 30 and 50, so 35, 42 and 49.*) **Is that the answer?** (*Yes.*) **How many teams of each number could there be?** (*5, 6, 7*)

4 **Which numbers need to be multiplied?** (*2953 and 6*) **How does this calculation need to be set out?** (*2953 with × 6 underneath.*) Let children work it out. (*17 718*)

5 **Which number is divided?** (*342*) **By how much?** (*6*) **How do we write this to solve it?** (*342 first then ÷ 6 alongside on the left. Answer is 57.*)

6 **How can we work this out without counting the squares?** (*6 × 6 = 36*) **How could we write this, using square number notation?** (6^2)

7 How can we work this out without counting the cubes, especially as we cannot see them all? (*cube number, 4 × 4 × 4*) **How could we write this using cube number notation?** (*4³ = 4 × 4 × 4 = 64*)

8 Which type of multiplication is this? (*long: 452 × 24*) **How can we write this to work it out?** (*452 with × 24 underneath; 10 848*)

Teaching activity 1b (20 minutes)

Solve problems involving multiplication, division, factors, multiples, square and cube numbers

- Use the cards cut from Resource 12: Multiplication and division word problems, separated into questions and operations. Help children to match the problems to the operations. As there is sometimes more than one way of solving a problem there may be some overlap, but try to match one operation card to each problem. Match the ones that are more obvious, then see how the others could fit (1D, 2H, 3E, 4B, 5C, 6F, 7G, 8A is the best fit).

- Shuffle the question cards and place them face down, then turn over the first one. Explain that the children are going to rewrite this problem so that the numbers or contexts are different, but the way of solving it is the same. Make simple changes, then solve the problem together. (See activity 1a for answers to the original questions and how to record them.) Here are some starters.

1 Change the class size to 24, 32, 28 and the context to a dance, gym or sports club. Ask: **What happens if the class size is a prime number?** (*Only the whole class or one-person groups are possible.*)

2 Change to 7-a-side, 11-a-side; groups of three for a dance competition; repacking packets of 6 biscuits into packets of 4 or 8 or any numbers that involve multiplication then division with remainders.

3 Change the numbers, from 20 to 40, 40 to 60; change the group size to 4, 5, 6, 8, 9, 11, 12.

4 Change the numbers and use a different context, for example: students in secondary school; cost of holidays. Keep the range to four-digit × one-digit.

5 Change to anything that needs a simple division. The children may have to deal with remainders, although they do not have to solve each one.

6 Change to a different sized square; change context to area if appropriate: use a square field 8 m by 8 m; what is its area?

7 Change to different sized cubes; change context to volume if appropriate: use a cube of 8 cm by 8 cm by 8 cm: what is its volume?

8 Change the numbers and context but stay with HTO × TO or ThHTO × TO. For example: **One computer costs £385. How much will 18 computers cost?** (*£6930*)

Unit 20: Solve problems involving addition, subtraction, multiplication and division and a combination of these, including understanding the meaning of the equals sign

Content domain reference: 5C8b

Prerequisites for learning

Use all four operations to solve problems
Know and use the = symbol

Learning outcomes

Balance what is on either side of an = sign
Solve multi-step word problems

Key vocabulary

Equals, equivalent, multi-step

Resources

Sets of digit cards 0–9; sets of cards showing +, –, ×, ÷ and =

Background knowledge

Children often misread multi-step problems, only partially completing the problem. They can be confused when there is more than one number on both sides of the equals sign.

Teaching activity 1a (15 minutes)

Balance what is on either side of an = sign

- Begin by asking what the = sign means in a familiar setting, such as $4 \times 5 = 20$. Ask: **Can you give me another pair of numbers that multiply to make 20?** Explain that you can write this as $4 \times 5 = 2 \times 10$ because both sides are equal to 20: 4×5 and 2×10 are equivalent calculations because they are both equal to 20.

- Write this example on a whiteboard: $3 \times 4 = 2 \times \square$. Ask: **What is the missing number?** (6) Repeat with other pairs of equivalent multiplications, changing the number represented by the box ($\square$) and its position: $4 \times \square = 8 \times 3$; $\square \times 3 = 2 \times 9$.

- Repeat with some additions and additions/subtractions, stressing that both sides must balance, so they must have the same value. Write: $24 + 6 = \square + 11$. Ask: **What number goes in the box?** (19) Say: **The left side equals 30, so the right side must also equal 30.** Write: $42 - \square = 25 + 16$. Say: **25 + 16 = 41 so 42 – $\square$ must equal 41. What goes in the box?** (1)

- Continue with more complicated calculations, continually highlighting that each side must have the same value and that the calculation in the brackets should be completed first: $(2 + 7) \times 4 = \square \times (8 + 4)$: $2 + 7 = 9$, $9 \times 4 = 36$; $\square \times 12 = 36$ so $\square = 3$.

Teaching activity 1b (15 minutes)

Balance what is on either side of an = sign

- Lay out the digit cards 4 and 5 with a gap between them for an operation card, then the equals sign. Then lay out the numbers 2 and 10 on the other side, with a gap between them: $5 \square 4 = 10 \square 2$. Explain that the calculations on both sides of the equals sign must have exactly the same value. Say: **They are equivalent calculations because they are both equal to the same amount.** Ask: **Which operation sign needs to go here to make both sides the same?** (×) Try each one in turn to check the results.

  ♦ Try +: this gives $5 + 4 = 10 + 2$ which implies that $9 = 12$, so it is not +.

  ♦ Try –: this gives $5 - 4 = 1$ but $10 - 2 = 8$.

  ♦ Try ×: it works.

  ♦ Try ÷: $5 \div 4$ does not have the same value as $10 \div 2 = 5$.

- The correct operation is multiplication (×).

- Repeat with other examples that use the same sign on both sides: $30 \square 6 = 20 \square 4$ (÷); $24 \square 6 = 19 \square 11$ (+). Then extend to using two different signs: $42 \square 1 = 25 \square 16$ (–, +); $42 \square 2 = 3 \square 7$ (÷, ×).

- Now write an 'answer' such as 24 and ask children to make up two different calculations connected by the equals sign, using two different operations, so that both sides make 24; for example: 6 × 4 = 32 − 8. Say: **Make as many different calculations as you can.**

- Finally, include some more complicated calculations with brackets: (2 ☐ 7) ☐ 4 = 3 ☐ (8 ☐ 4). Discuss how 2 ☐ 7 is unlikely to be ÷ or, so try + and ×. There could be more than one solution and children can swap the cards around as much as they like, but they do need to use a systematic approach: keep one or more the same and swap the others. Stress that these result in a whole number on either side of the = (+, × = ×, +).

Teaching activity 2a (20 minutes)

Use all four operations to solve problems

- Consider together the word problem:
  Twins Ruben and Finn saved **£35** *each, every* **month** *for a* **year**. *How many more months do they need to have a combined total of* **£2000?**
  Ask: **What are the keys words and numbers in this problem?** Highlight them together.
  Ask: **What do we need to know?** (*There are 12 months in a year.*) **What do we need to work out first?** (*either 35 × 12 doubled or 70 × 12 = 840*) **What is next?** (*2000 − 840 = 1160*)
  What next? (*1160 ÷ 70*) Say: **This will have the same answer as 116 ÷ 7.** Ask: **Can you tell me why?** (*dividing both numbers by 10; think about 240 ÷ 40 = 24 ÷ 4*) Say: **116 ÷ 7 = 16 remainder 4. Is the answer 16? Will they have enough in another 16 months or do they need to save for another month?** (*They need to save for another month as this will take them over the £2000, so the answer is 17.*)

- Ensure that you write each calculation in the correct order and recap how to work each of them out, using the methods taught in these units. Discuss the fact that sometimes division involves remainders and the answer requires us to round up or down, depending on the context.

Teaching activity 2b (20 minutes)

Use all four operations to solve problems

- Consider together the word problem:
  Twins Ruben and Finn saved **£35** *each, every* **month** *for a* **year**. *How many more months do they need to have a combined total of* **£2000?**
  Explain that it is sometimes easier to sketch the problem or draw a table to solve it. Instead of straight calculations children can use a trial and improvement model. This one uses doubling and other mental strategies.

Ruben   Finn   = £70 each month in total

Month	Total (£)
1	70
12	12 × 70 = 840
24	840 × 2 = 1680
30	70 × 30 = 2100
29	**2100 − 70 = 2030**
28	2 100 − 140 = 1960

- Each month they save £70 altogether.

- In a year (12 months) they have saved £840, not yet half of what they need, so double it for 2 years (24 months): they have £1680, still not enough; try 30 months, which will be £2100, a little too much.

- 29 months is just over £2000, but 28 months is not quite enough, so they need to save for 29 months. They have already saved for 12 months, so they need to save for another 17 months to have £2000.

Unit 21: Solve problems involving multiplication and division, including scaling by simple fractions and problems involving simple rates

Content domain reference: 5C8c

Prerequisites for learning

Divide whole numbers by numbers from 2 to 10
Find fractions of amounts by counting objects

Key vocabulary

Fraction of, equivalent, bar model

Resources

Resource 9: Bar models; Resource 13: Fractions of 24; slips of paper (18 per pair); small counters; coloured pencils

Learning outcomes

Find fractions of amounts
Solve problems involving simple rates

Background knowledge

Children need to make the connection between division and the denominator of a fraction. They need to have experience of the different forms questions may take, for example: '$\frac{1}{3}$ of' is the same as '$\frac{1}{3} \times$' and is the same as $\div 3$: $\frac{1}{3}$ of $24 = \frac{1}{3} \times 24 = 24 \div 3$.

Teaching activity 1a (15 minutes)

Find fractions of amounts

- Let the children work in pairs. Ask one child in each pair to write $\div 2$, $\div 3$, $\div 4$, $\div 5$, $\div 6$, $\div 7$, $\div 8$, $\div 9$ and $\div 10$ on separate pieces of paper and then mix them up. Ask the other child in the pair, at the same time, to write $\frac{1}{2}$, $\frac{1}{3}$, $\frac{1}{4}$, $\frac{1}{5}$, $\frac{1}{6}$, $\frac{1}{7}$, $\frac{1}{8}$, $\frac{1}{9}$, $\frac{1}{10}$ on separate pieces of paper. Then ask them, working as a pair, to see how quickly they can match the fraction with the division.

- Establish that finding a fraction of an amount is the same as **d**ividing that amount by the **d**enominator (both **d**s).

- Use Resource 9: Bar models and small counters. Start with thirds. Say: **This bar is split into 3 equal parts, so each is $\frac{1}{3}$. We are going to work out $\frac{1}{3}$ of 12.** Ask children to divide their 12 counters equally among the three sections of the first three-part bar. Ask: **How many counters are there in each third?** (*4*) Say: **We say that $\frac{1}{3}$ of 12 is 4. What is 12 divided by 3?** (*It is 4, just the same.*) Repeat with another example involving thirds, such as $\frac{1}{3}$ of 15 or $\frac{1}{3}$ of 18. Establish that $\frac{1}{3}$ of $15 = 15 \div 3 = 5$. Ask: **What is $\frac{2}{3}$ of 12?** (*4 + 4 = 8 or 2 × 4 = 8*) Say: **$\frac{2}{3}$ is simply two thirds. One third = 4 so two thirds must equal 8.**

- Repeat with quarters, fifths and sixths, as necessary, recording each as a fraction and identifying the equivalent division until children can confidently find a unit fraction of quantity without using counters.

- Repeat with finding non-unit fractions ($\frac{2}{3}$, $\frac{3}{4}$, $\frac{3}{5}$ …). Many children will begin to understand that they need to divide by the denominator then multiply by the numerator. Use counters or pencil marks in the bars on Resource 9: Bar models to provide visual support as necessary.

Teaching activity 1b (20 minutes)

Find fractions of amounts

- Distribute copies of Resource 13: Fractions of 24. Ask children to divide the first grid of 24 into half and to count the cells in that half. Ask them to write underneath: $\frac{1}{2}$ of $24 = 12$ and $24 \div 2 = 12$, to make the link between division and the denominator of the fraction.

- Show children how to use the columns to split the next grid into thirds: each third will be two columns wide. Ask: **How many cells are in this third?** (*8*) **How can we record this as a fraction of 24?** ($\frac{1}{3}$ *of 24*) **How can we record this as a division?** (*24 ÷ 3 = 8*)

- Repeat for quarters, sixths, twelfths and twenty-fourths. Discuss the best way to split the grid. ($\frac{1}{4}$ = *1 row*; $\frac{1}{6}$ = *1 column*, $\frac{1}{12}$ = $\frac{1}{2}$ *a column*, $\frac{1}{24}$ = *1 cell*)

- Now look at some non-unit fractions, such as $\frac{3}{4}$ of 24, starting with the grid divided into quarters. Say: $\frac{3}{4}$ **is three of these quarters, so how many cells do we need to colour?** (*18*) Show children how to record this as a fraction statement: $\frac{3}{4}$ of 24 = 18. Discuss how they found $\frac{1}{4}$ then multiplied it by 3. Say: **Divide by the denominator, multiply by the numerator.** Record this as a number sentence: 24 ÷ 4 × 3 = 18. Repeat for other fractions, using the unit fraction diagram for support. Calculate amounts such as: $\frac{2}{3}$ of 24; $\frac{5}{6}$ of 24, $\frac{5}{12}$ of 24; $\frac{7}{24}$ of 24. Record each one as a fraction of 24 and as a number sentence involving division and multiplication: $\frac{2}{3}$ of 24 = 16, 24 ÷ 3 × 2 = 16.

Teaching activity 2a (15 minutes)

Solve problems involving simple rates

- Ask children this question: **I bought three books, each costing the same amount. I paid £15. How much did each book cost?** Ask: **What do we need to do to solve this problem?** (*Divide 15 by 3.*) Ask a child to write this as a number sentence: 15 ÷ 3 = 5.

- Now set the problem: **Tom bought five identical notebooks for £11. What was the cost of each notebook?** Discuss how this is more difficult to solve because 11 is not a multiple of 5. 11 ÷ 5 = 2 remainder 1. Stating the remainder as whole number does not help, but some children may be able to see that this is $\frac{1}{5}$ of £1 (100p) = 20p.

- Ask children to suggest other 'multi-buy' questions. Ask them to think of examples with a whole number solution (when the number of items is a factor of the total cost).

- Now ask this question: **Four pens cost £12. How much will five pens cost?** Explain that they need to find the cost of one pen first, by dividing, then multiply to find the cost of five. Record the working, using number sentences: 12 ÷ 4 = 3; 3 × 5 = 15. Five pens cost £15. Repeat with another example: **Nine pizzas costs £36, how much will four pizzas cost?** (*36 ÷ 9 × 4 = £16*)

- Consider examples such as: **I paid £42 for eight diaries, so how much would I pay for four?** (*Simply halve £42.*) **I paid £20 for six pens. How much would I pay for 12?** (*Just double the total.*) This saves having to find the cost of one, especially as the numbers involved will produce remainders.

Teaching activity 2b (15 minutes)

Solve problems involving simple rates

- Use Resource 9: Bar models. Start with a word problem: **I bought three books, each costing the same amount. I paid £15. How much did each book cost?** Explain that they will use a bar model to illustrate this.

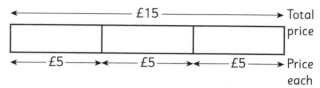

Say: **One book will cost £5, two books will cost £10 because £15 ÷ 3 = £5; £5 × 2 = £10.**

- Repeat for four items, again using the four-bar model: **I paid £24 for four calculators, how much would I pay for one calculator? For two calculators? For three calculators? How can we write this using number sentences?** (*24 ÷ 4 = £6; £6 × 2 = £12; £6 × 3 = £18*)

- Repeat for other multiple buys, from 5 to 10, keeping the total cost as a multiple of the number of items bought. Use the relevant bar-model on Resource 9 and find the cost of one, then multiple items, recording the calculations.

- Finally, give some examples such as: **I bought seven pencil cases for £35, how much would I pay for five pencil cases?** Show that this can be recorded as £35 ÷ 7 × 5 = £25. Use bar models for visual support as necessary.

Unit 22: Identify, name and write equivalent fractions of a given fraction, represented visually, including tenths and hundredths

Content domain reference: 5F2b

Prerequisites for learning

Find fractions of numbers
Split a diagram into a given fraction

Key vocabulary

Numerator, denominator, equivalent fraction, tenths, hundredths, ≡, factor, multiple

Learning outcomes

Understand equivalence in fractions

Resources

Resource 13: Fractions of 24; Resource 14: Hundredths; squared paper

Background knowledge

Find fractions of amounts
Split a diagram to show a given fraction

Teaching activity 1a (40 minutes)

Understand equivalence in fractions

- Distribute Resource 13: Fractions of 24 or squared paper. Ask children to split the first 6 by 4 diagram into halves and to shade one half. They write $\frac{1}{2}$ under the diagram.

- Ask them to split the next diagram into quarters **as halves of a half** and shade **2** quarters, writing $\frac{2}{4}$ underneath. They then split the third diagram into eighths by repeated halving, and shade **4 eighths,** writing $\frac{4}{8}$ underneath.

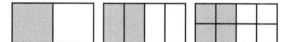

- Write $\frac{1}{2} \equiv \frac{2}{4} \equiv \frac{4}{8}$ Explain that the symbol ≡ means 'is equivalent to', which simply means 'is equal to'. Say: **Two quarters make the same fraction as a half. $\frac{2}{4}$ and $\frac{1}{2}$ are equivalent fractions. Who can predict the next equivalent fraction in this sequence?** (*The numerators and the denominators are both doubling, so $\frac{4}{8}$, $\frac{8}{16}$, then $\frac{16}{32}$.*) Explain that we cannot show this last fraction on a grid of 24 squares.

- Say: **Look again at these fractions that are equivalent to one half.** Ask: **What is the relationship between the numerator and the denominator in each of them?** (*The numerator is half the denominator.*) **How many tenths would be equivalent to a half?** (*5, as 5 is half of 10.*) **How many hundredths would be equivalent to a half?** (*50, as 50 is half of 100.*)

- Ask each child to say another fraction that is equivalent to a half. If they struggle, emphasise that the numerator must be half of the denominator or the denominator must be double the numerator. Give children a numerator and ask for the denominator; give them a denominator and ask for the numerator.

- Help the children to split the next three 24-grids into thirds, sixths and twelfths, shading $\frac{1}{3}$, $\frac{2}{6}$ and $\frac{4}{12}$. Write the equivalence sentence: $\frac{1}{3} \equiv \frac{2}{6} \equiv \frac{4}{12}$. Ask: **Can you think of another fraction that is equivalent to a third?** ($\frac{5}{15}$) **What is the relationship between the numerator and the denominator?** (*It is a third.*) **Can we find a fraction of tenths or hundredths that is equivalent to a third?** (*No, as 3 is not a factor of 10 or 100.*)

- Repeat, but this time with children shading $\frac{2}{3}$ of each diagram to demonstrate $\frac{2}{3} \equiv \frac{4}{6} \equiv \frac{8}{12}$. Say: **The relationship between numerator and denominator is not so easy to spot, but the numerator is $\frac{2}{3}$ of the denominator. Can you think of another fraction that is equivalent to $\frac{2}{3}$?** ($\frac{10}{15}$)

- Use the next two sets of three 24-grids to show $\frac{1}{4} \equiv \frac{2}{8} \equiv \frac{4}{16}$ and $\frac{3}{4} \equiv \frac{6}{8} \equiv \frac{12}{16}$. Then ask the children to think of other fractions that are equivalent to $\frac{1}{4}$ and $\frac{3}{4}$. Ask them to describe the relationship between the numerator and denominator. Then ask: **Can you find a fraction of tenths that equivalent to $\frac{1}{4}$?** *(No, as 4 is not a factor of 10.)* **Can you find a fraction of hundredths that equivalent to $\frac{1}{4}$?** *(Yes: 4 is a factor of 100, so $\frac{25}{100} \equiv \frac{1}{4}$ and $\frac{75}{100} \equiv \frac{3}{4}$.)*

- If there is time, explore making equivalent fractions by multiplying the numerator and denominator by the same number.

Teaching activity 1b (40 minutes)

Understand equivalence in fractions

- Distribute Resource 14: Hundredths. Explain that there are 100 small squares in each diagram. There are 10 rows of 10 squares, $10 \times 10 = 100$, which is a square number (Unit 18). Ask children to split the first hundred square into halves and to write $\frac{1}{2}$ underneath. Ask: **How many small squares are shaded?** *(5 rows of 10 or 5 columns of 10 = 5, 50 out of 100)* Say: **This is the fraction $\frac{50}{100}$.** Write '$\equiv \frac{50}{100}$' next to the $\frac{1}{2}$, reminding children that the symbol $\equiv$ means 'is equivalent to'. Say: **We could just use an equals sign instead of $\equiv$.**

- Ask children to split the next 100 square into quarters and write $\frac{1}{4}$ underneath. Ask: **How many small squares are shaded?** *(5 rows of 5 or 5 columns of 5 = 25, 25 out of 100)* **What is this as a fraction?** *($\frac{25}{100}$)* Ask children to write $\frac{25}{100} \equiv \frac{1}{4}$ under the diagram.

- Repeat, with children shading $\frac{3}{4} \equiv \frac{75}{100}$; $\frac{1}{10} \equiv \frac{10}{100}$; $\frac{1}{5} \equiv \frac{20}{100}$; $\frac{7}{10} \equiv \frac{70}{100}$.

- Ask children to draw squares of different dimensions (3×3, 4×4, 5×5, 6×6). They use them to explore equivalent fractions by shading whole cells. They should record each one, using the $\equiv$ symbol. Discuss which fractions they can find and which they cannot: an odd number of cells cannot be split into fractions with an even denominator. Remind them that using factors of the number of cells is a useful strategy and that these diagrams all show square numbers.

- Discuss the large number of equivalent fractions and explain that they have found only some of the fractions equivalent to $\frac{1}{2}$, $\frac{1}{3}$, $\frac{1}{4}$... Tell them that equivalent fractions can also be found by multiplying the numerator and denominator by the same number:

  ◆ $\frac{1}{5}$: multiplying both numbers by 7 gives $\frac{7}{35}$, $\frac{7}{35}$ is equivalent to $\frac{1}{5}$

  ◆ $\frac{2}{3}$: multiplying both numbers by 10 gives $\frac{20}{30}$, $\frac{20}{30}$ is equivalent to $\frac{2}{3}$.

$$\begin{array}{c} \times 7 \\ \dfrac{1}{5} = \dfrac{7}{35} \\ \times 7 \end{array} \qquad \begin{array}{c} \times 10 \\ \dfrac{2}{3} = \dfrac{20}{30} \\ \times 10 \end{array}$$

Unit 23: Compare and order fractions whose denominators are all multiples of the same number

Content domain reference: 5F3

Prerequisites for learning

Multiply mentally by one-digit numbers

Find equivalent fractions

Key vocabulary

Numerator, denominator, equivalent fraction, compare, order

Learning outcomes

Compare and order fractions with related denominators

Resources

Resource 15: Fraction wall; mini-whiteboards; rulers

Background knowledge

Although most children will understand that the larger the denominator, the smaller the fractional part, comparing non-unit fractions (numerator ≠ 1) will often cause confusion. For example, $\frac{10}{11}$ is larger than $\frac{9}{10}$, even though elevenths are smaller than tenths. Understanding that different fractions can be equivalent (Unit 22), and knowing how to use this equivalence to compare fractions, is the basis for adding, subtracting, multiplying and dividing fractions effectively. Having a clear understanding of whether a fraction is more than, less than or equal to a half is a very useful skill.

Teaching activity 1a (30 minutes)

Compare and order fractions with related denominators

- Use Resource 15: Fraction wall, to play 'More than, less than, equal to $\frac{1}{2}$'. Write a fraction on a whiteboard. The children respond by writing >, < or = on their mini-whiteboards, referring to the length of the bars representing fractions in the resource sheet and using a ruler to help them follow the halfway mark.

- Start with some unit fractions (with numerator 1), establishing that they are all less than $\frac{1}{2}$. Then try, in any order: $\frac{2}{3}, \frac{2}{4}, \frac{4}{5}, \frac{7}{10}$ ($>\frac{1}{2}$); $\frac{2}{5}, \frac{3}{8}, \frac{4}{9}, \frac{5}{12}$ ($<\frac{1}{2}$) and $\frac{2}{4}, \frac{5}{10}, \frac{6}{12}$ ($=\frac{1}{2}$).

- Ask: **How can we tell, without using the bars, whether a fraction is more than, less than or equal to $\frac{1}{2}$?** (*If the numerator is more than half of denominator it is more than $\frac{1}{2}$; if the numerator is less than half of denominator it is less than $\frac{1}{2}$; if the numerator is exactly half of denominator it equals $\frac{1}{2}$.*) Test children's understanding by writing fractions that are not shown on the bars.

- Use Resource 15: Fraction wall to demonstrate which fraction is larger, $\frac{2}{3}$ or $\frac{5}{6}$ ($\frac{5}{6}$ is a longer bar so is larger). First, ask: **How many sixths are there in the fraction that is equivalent to $\frac{2}{3}$?** Use a ruler to demonstrate how to move down from the thirds to the sixths, to show that $\frac{2}{3}$ is the same length as, and therefore equivalent to, $\frac{4}{6}$. Then elicit that $\frac{4}{6} < \frac{5}{6}$ so $\frac{2}{3} < \frac{5}{6}$.

- Now write pairs of fractions on a whiteboard and ask the children to write down the fraction with the larger value. Use these fraction families: $\frac{1}{2}, \frac{1}{4}, \frac{1}{8}$; $\frac{1}{2}, \frac{1}{3}, \frac{1}{6}, \frac{1}{12}$; $\frac{1}{2}, \frac{1}{5}, \frac{1}{10}$. Provide plenty of practice, with the aim that children will be able to compare the fractions without using the length of the fraction bars.

- Finally, use the bars in Resource 15: Fraction wall, or what children know about equivalence, to order three fractions, from smallest to largest. Start with groups of fractions comprising $\frac{1}{2}$, one fraction **less** than $\frac{1}{2}$ and one fraction **more** than $\frac{1}{2}$, such as: $\frac{1}{2}, \frac{1}{3}, \frac{2}{3}$; $\frac{3}{4}, \frac{5}{8}, \frac{1}{2}$; $\frac{1}{3}, \frac{4}{5}, \frac{5}{10}$; $\frac{5}{12}, \frac{5}{6}, \frac{1}{2}$.

- Continue with more sets, taken from the fraction families above. Demonstrate the first one, using the length of the fraction bars and then equivalence. For example, to compare $\frac{2}{3}, \frac{5}{6}, \frac{7}{12}$, using the bars, say: **$\frac{7}{12}$ is the shortest, $\frac{2}{3}$ next and $\frac{5}{6}$ is the largest.** Then, using equivalence: **$\frac{7}{12}$ has the highest number as its denominator, so we need to change $\frac{2}{3}$ and $\frac{5}{6}$ into twelfths. $\frac{2}{3} \equiv \frac{8}{12}$; $\frac{5}{6} \equiv \frac{10}{12}$. Now it is easy to order them.**

Teaching activity 1b (30 minutes)

Compare and order fractions with related denominators

- Play 'More than, less than, equal to $\frac{1}{2}$'. Write a fraction on a whiteboard. The children respond by writing >, < or = on their mini-whiteboards. If they are struggling, discuss how they can tell, again by comparing the numerator to the denominator. (*Half the numerator is more than half the denominator, >; < half the numerator is less than half of denominator, <; half the numerator is exactly half of denominator, =.*) Use a variety of fractions, including some challenging ones such as $\frac{49}{100}$, $\frac{25}{50}$, $\frac{42}{80}$.

- Explain that the children are now going to compare two fractions by changing one of the fractions into an equivalent fraction, with the same denominator as the other fraction. To do this, they will use multiplication. Demonstrate, asking: **Which is larger $\frac{2}{3}$ or $\frac{5}{6}$?** Say: **We can change the smaller denominator into the larger denominator by doubling both the numerator and denominator:** $\frac{2}{3} = \frac{4}{6}$. (*See unit 22.*) **We know that $\frac{4}{6} < \frac{5}{6}$ so $\frac{2}{3} < \frac{5}{6}$. $\frac{5}{6}$ is the larger fraction.**

- Write pairs of fractions on a whiteboard. Let children choose a pair and work out which is the larger. Encourage them to show their working, showing which fraction they have changed to prove which is the larger. Use these fraction families: $\frac{1}{2}$, $\frac{1}{4}$, $\frac{1}{8}$; $\frac{1}{2}$, $\frac{1}{3}$, $\frac{1}{6}$, $\frac{1}{12}$; $\frac{1}{2}$, $\frac{1}{5}$, $\frac{1}{10}$.

- Finally, use equivalence to order three fractions, **smallest to largest**. Encourage children to use their knowledge of <, > = $\frac{1}{2}$.

- Continue with more sets, still using the fraction families above. Demonstrating the first one, $\frac{2}{3}$ using equivalence. For example, to compare $\frac{2}{3}$, $\frac{5}{6}$, $\frac{7}{12}$, say: $\frac{7}{12}$ **has the highest number as its denominator, so we need to change $\frac{2}{3}$ and $\frac{5}{6}$ into twelfths.** Ask: **To change $\frac{2}{3}$, what do we need to multiply the 3 by to get a denominator of 12?** (*4 × 3 = 12*) Say: **Then we need to multiply the numerator by 4 as well.** Write: $\frac{2}{3} \equiv \frac{8}{12}$. Say: **To change $\frac{5}{6}$, what do we need to multiply the 6 by to get a denominator of 12?** (*by 2 or double both numbers*) Write: $\frac{5}{6} \equiv \frac{10}{12}$. Say: **Now it is easy to order them. We have $\frac{7}{12}$, $\frac{8}{12}$, $\frac{10}{12}$, so the order is $\frac{7}{12}$, $\frac{2}{3}$, $\frac{5}{6}$.**

- Let children try ordering sets such as: $\frac{3}{4}$, $\frac{5}{8}$, $\frac{1}{4}$ and $\frac{3}{5}$, $\frac{3}{10}$, $\frac{2}{5}$ by themselves.

- **Challenge**: Ask children to order $\frac{2}{3}$, $\frac{5}{9}$, $\frac{3}{6}$. Say: **We can change $\frac{2}{3}$ into ninths, but we cannot change sixths into ninths. What do you notice about the sixths?** ($\frac{3}{6} \equiv \frac{1}{2}$) **We then need to change $\frac{2}{3}$ into ninths, then decide if they are more or less than $\frac{1}{2}$: $\frac{2}{3} = \frac{6}{9}$; $\frac{6}{9} > \frac{5}{9}$, and both $\frac{5}{9}$ and $\frac{6}{9}$ are more than $\frac{1}{2}$, so the order is $\frac{3}{6}$, $\frac{5}{9}$, $\frac{2}{3}$.**

- Children try ordering $\frac{4}{8}$, $\frac{3}{4}$ and $\frac{5}{12}$ in the same way.

Unit 24: Recognise mixed numbers and improper fractions and convert from one form to the other; write mathematical statements > 1 as a mixed number

Content domain reference: 5F2a

Prerequisites for learning

Recognise and use equivalence in fractions

Learning outcomes

Convert between mixed numbers and improper fractions

Add fractions of the same denominator, giving answers > 1 as mixed numbers

Key vocabulary

Proper fraction, improper fraction, mixed number, convert, numerator, denominator

Resources

Resource 16: Fraction circles; mini-whiteboards; coloured pencils

Background knowledge

In **proper fractions** the numerator is less than the denominator, so the value is less than one. They do not have to be in their simplest form. Proper fractions are also known as vulgar fractions.

Fractions with value greater than one are called **improper** fractions. The numerator is greater than the denominator, so the fraction is 'top heavy', for example, $\frac{5}{3}$.

A **mixed number** is made up of a whole number and a proper fraction, for example, $1\frac{1}{8}$, $2\frac{3}{4}$.

Children need to be able to convert between mixed numbers and improper fractions before they can successfully calculate with fractions. Adding the numerators **and** the denominators is a common error when children add fractions.

Teaching activity 1a (15 minutes)

Convert between mixed numbers and improper fractions

- Write some proper fractions on the board: $\frac{2}{3}, \frac{5}{6}, \frac{1}{2}, \frac{1}{3}, \frac{1}{4}$. Ask: **What do you notice about the numerator? Is it larger or smaller than the denominator?** (*smaller*) Explain that these fractions are called proper fractions: fractions in which the numerator is smaller than the denominator. Ask: **What do you think an improper fraction could be?** (*One in which the numerator is larger than the denominator.*) Write some examples on a whiteboard: $\frac{5}{2}, \frac{7}{3}, \frac{6}{4}$. Ask children to suggest some more, checking each one together carefully to ensure that its numerator is larger than its denominator and writing them on a different board.

- Ask: **What do you think a mixed number could be?** Write some examples on a separate whiteboard: $2\frac{1}{2}, 4\frac{1}{4}, 6\frac{1}{3}$. Tell the children that it is a number with a whole-number part and a fractional part.

- Explain that they are going to convert or change between mixed numbers (point to the board of mixed numbers) and improper fractions (point to the board of improper fractions).

- Give the children Resource 16: Fraction circles. On the halves row, ask them to shade in two whole circles and half of the third one. Ask: **How many whole circles and how many half circles are shaded? How could we write this as a mixed number?** ($2\frac{1}{2}$) **How many half circles are shaded?** (5) **How could we write this as a fraction of halves?** ($\frac{5}{2}$, which is an improper fraction. $2\frac{1}{2}$ is the same as $\frac{5}{2}$. 1 whole can be written also as $\frac{2}{2}$, so $\frac{2}{2} + \frac{2}{2} + \frac{1}{2} = \frac{5}{2}$.)

- Repeat with thirds: shade three whole circles and two sections of the fourth circle. Ask children to write this as a mixed number and as an improper fraction. Ask: **How many whole parts?** (3) **What fraction is shaded in this fourth circle?** ($\frac{2}{3}$) **How can we write this as a mixed number?** ($3\frac{2}{3}$) **How many thirds are shaded altogether?** (11) **How can we write this as an improper fraction of thirds?** ($\frac{11}{3}$)

- Repeat for quarters ($1\frac{3}{4} = \frac{7}{4}$), fifths ($3\frac{4}{5} = \frac{19}{5}$) and tenths ($2\frac{7}{10} = \frac{27}{10}$), asking the same key questions as above and writing each diagram as a mixed number and as an improper fraction.

Teaching activity 1b (15 minutes)

Convert between mixed numbers and improper fractions

- Explain the different types of fraction: proper, improper and mixed numbers.

- Explain that we can use multiplication and addition to convert a mixed number to an improper fraction, and division to convert an improper fraction to a mixed number.

- Explain how to convert a mixed number to an improper fraction. To find the numerator, multiply the whole number part by the denominator of the fraction, then add this to the original numerator. The denominator remains the same.
  $2\frac{3}{4}$: $2 \times 4 + 3 = 11$ so $\frac{11}{4} = 2\frac{3}{4}$;
  $4\frac{2}{5}$: $4 \times 5 + 2 = 22$, so $4\frac{2}{5} = \frac{22}{5}$

- Practise some more examples. You can invite children to suggest the mixed number, but the fraction part should be in its simplest form for best results.

- Explain how to convert an improper fraction to a mixed number, using division and remainders: to find the whole number, divide the numerator by the denominator, the remainder is the numerator of the fractional part and the denominator stays the same:
  $\frac{7}{4}$: $7 \div 4 = 1$ r 3, so $\frac{7}{4} = 1\frac{3}{4}$
  $\frac{22}{5}$: $22 \div 5 = 4$ r 2, so $\frac{55}{5} = 4\frac{2}{5}$

- Practise some more examples, keeping the numerators no greater than 10.

- Initially, some children may need the support of quick sketches, using Resource 16: Fraction circles (see activity 1a.)

Teaching activity 2a (15 minutes)

Add fractions of the same denominator, giving answers > 1 as mixed numbers

- Write $\frac{1}{2} + \frac{1}{2}$ on a whiteboard. Children may already be able to give the answer. Using Resource 16: Fraction circles, show this on the first halves circle, with each half coloured a different colour – two halves make one whole.

- Use thirds to add $\frac{2}{3} + \frac{2}{3}$. Shade $\frac{2}{3}$ of the first circle in one colour and then the last third plus one of the thirds of the second circle a different colour. Avoid colouring $\frac{2}{3}$ of the first and then $\frac{2}{3}$ of the second as children might interpret this as 4 out of 6, or $\frac{4}{6}$ which equals $\frac{2}{3}$ which is not the correct answer for $\frac{2}{3} + \frac{2}{3}$. Ask: **How many thirds have we coloured?** (*4 thirds*) **How do we write $\frac{2}{3} + \frac{2}{3}$ as an improper fraction?** ($\frac{4}{3}$) **How can we write it as a mixed number?** ($1\frac{1}{3}$)

- Use quarters to calculate $\frac{3}{4} + \frac{3}{4}$, fifths for $\frac{4}{5} + \frac{3}{5}$ and tenths for $\frac{7}{10} + \frac{9}{10}$. Repeat the previous question sequence, recording the answers as improper fractions and as mixed numbers.

- Try $\frac{5}{6} + \frac{5}{6}$ without the support of diagrams.

Teaching activity 2b (15 minutes)

Add fractions of the same denominator, giving answers > 1 as mixed numbers

- Ask children to remind each other how to add fractions that have the same denominator (*Add the numerators; keep the denominator the same.*)

- Show an example giving a sum of less than 1, as a reminder: $\frac{2}{5} + \frac{1}{5} = \frac{3}{5}$. Stress that only the numerators are added.

- Now consider $\frac{5}{8} + \frac{3}{8} = \frac{8}{8}$. Ask: **What are eight-eighths?** (*one whole*) $\frac{5}{8} + \frac{3}{8} = 1$

- Work through $\frac{5}{8} + \frac{7}{8} = \frac{12}{8}$. Ask: **How can we write this as a mixed number?** (*Use division.*) $\frac{12}{8} = 1\frac{4}{8}$. Ask: **What is $\frac{4}{8}$ equivalent to?** ($\frac{1}{2}$) Say: **So $\frac{5}{8} + \frac{7}{8} = 1\frac{1}{2}$ is the best answer.**

- Complete a few more examples, including some examples in which the fraction in the answer can be simplified: $\frac{5}{9} + \frac{7}{9} = \frac{12}{9} = 1\frac{3}{9} = 1\frac{1}{3}$; $\frac{7}{10} + \frac{9}{10} = \frac{16}{10} = 1\frac{6}{10} = 1\frac{3}{5}$.

Unit 25: Add and subtract fractions with the same denominator, and denominators that are multiples of the same number

Content domain reference: 5F4

Prerequisites for learning

Convert between mixed numbers and improper fractions

Find fractions of numbers

Add and subtract fractions with the same denominator

Learning outcomes

Add and subtract fractions with related denominators

Key vocabulary

Numerator, denominator, mixed number, improper fraction, equivalent fraction, simplify, convert

Resources

Resource 7: Multiplication grid; Resource 15: Fraction wall; squared paper

Background knowledge

If you need to add or subtract fractions, they must have the same denominator. This means that one or more of the fractions may need to be converted to an equivalent fraction, with the same denominator as the other(s). If the fractions to be added are in their simplest forms, convert the fraction with the smaller denominator.

A common error is to add or subtract the numerators and the denominators. Address this by drawing diagrams or looking at some examples to show that this does not work: $\frac{1}{4} + \frac{1}{2}$ cannot be $\frac{2}{6}$ (adding the numerators, adding the denominators) as $\frac{2}{6}$ is less than $\frac{1}{2}$. Since $\frac{1}{4}$ has been added to $\frac{1}{2}$ the answer must be more than $\frac{1}{2}$.

Teaching activity 1a (20 minutes)

Add and subtract fractions with related denominators

- Revise adding fractions with the same denominator (Unit 24), converting any improper fractions to mixed numbers. Stress that the denominator stays the same:
$\frac{4}{5} + \frac{3}{5} = \frac{7}{5} (= 1\frac{2}{5}); \frac{9}{10} + \frac{9}{10} = \frac{18}{10} (= 1\frac{8}{10} = 1\frac{4}{5})$

- Revise subtracting fractions with the same denominator from Year 4, emphasising that only the numerators are subtracted; again, the denominator stays the same:
$\frac{5}{8} - \frac{3}{8} = \frac{2}{8} (= \frac{1}{4})$

- Explain that fractions can only be added or subtracted if they have the same denominator. If they have different denominators, we need to convert one or both, using equivalent fractions. Show children Resource 7: Multiplication grid. Ask: **Did you know that this multiplication grid shows equivalent fractions?** Point out how the first two rows, with row 1 as the numerators and row 2 as the denominators, show the fractions equivalent to $\frac{1}{2}$. The first and third rows

show the fractions equivalent to $\frac{1}{3}$. The fourth and fifth rows show $\frac{4}{5}$ and its equivalents. Ask: **Which rows show the equivalents to $\frac{2}{7}, \frac{5}{9}, \frac{2}{4}$?** (*rows 2 and 7, rows 5 and 9, rows 3 and 4*)

- Ask: **Can you find the fraction $\frac{24}{30}$ in the same column?** (*column 6*) **Which fraction is made by the first two numbers in those rows?** ($\frac{4}{5}$) Say: **We have simplified the fraction: $\frac{24}{30}$ = $\frac{4}{5}$.** Repeat with $\frac{45}{63}$. Ask: **Which column has both numbers?** (*column 9*) Trace along to the first column to simplify the fraction to $\frac{5}{7}$. Use the resource as a support where necessary.

- Demonstrate additions and subtractions in which one denominator is a multiple of the other, showing how to use equivalent fractions. Simplify the answer, converting to a mixed number where necessary. Use supportive prompting to encourage the children to do the work.

  ◆ $\frac{1}{4} + \frac{1}{2}$: Say: **We need to convert the fraction with the smaller denominator. Which fraction should we convert?** (*Convert half into quarters: $\frac{2}{4} + \frac{1}{4} = \frac{3}{4}$*)

◆ $\frac{9}{10} - \frac{2}{5}$: Ask: **Do we need to convert tenths into fifths or fifths into tenths? Which is the smaller denominator?** ($\frac{2}{5}$: $\frac{9}{10} - \frac{4}{10} = \frac{5}{10}$) **What is $\frac{5}{10}$ equivalent to?** ($\frac{1}{2}$)

◆ $\frac{2}{3} + \frac{5}{9} = \frac{6}{9} + \frac{5}{9} = \frac{11}{9}$ Ask: **What is $\frac{11}{9}$ as a mixed number?** ($1\frac{2}{9}$)

◆ $\frac{3}{4} - \frac{5}{12} = \frac{9}{12} - \frac{5}{12} = \frac{4}{12}$ Ask: **Can $\frac{4}{12}$ be simplified?** (*Yes, to $\frac{1}{3}$.*)

- Extend by considering some examples in which both denominators must be changed to equivalent fractions with a common denominator.

◆ $\frac{1}{6} + \frac{7}{9}$: Ask: **What number is a multiple of 6 and of 9?** (*18*) Say: **We need to convert both fractions into eighteenths: start with $\frac{1}{6}$: 3 × 6 = 18, so multiply numerator and denominator by 3: $\frac{3}{18}$. Then deal with $\frac{7}{9}$: 2 × 9 = 18, so multiply numerator and denominator by 2: $\frac{14}{18}$, therefore $\frac{3}{18} + \frac{14}{18} = \frac{17}{13}$.**

◆ $\frac{3}{4} - \frac{1}{3}$: Say: **3 and 4 are both factors of 12; 12 is a multiple of 3 and of 4, so we convert to twelfths: $\frac{9}{12} - \frac{4}{12} = \frac{5}{12}$.**

Teaching activity 1b (20 minutes)

Add and subtract fractions with related denominators

- Revise adding and subtracting fractions with the same denominator, using diagrams drawn on squared paper to support where necessary. Refer back to Unit 24, activities 2a and 2b.

- Revise adding fractions with the same denominator (Unit 24), converting any improper fractions to mixed numbers:

$$\frac{4}{5} + \frac{3}{5} = \frac{7}{5} = 1\frac{2}{5}; \frac{9}{10} + \frac{9}{10} = \frac{18}{10} = 1\frac{8}{10} = 1\frac{4}{5}$$

- Revise subtracting fractions with the same denominator from Year 4, emphasising that only the numerators are subtracted; the denominators stay the same: $\frac{5}{8} - \frac{3}{8} = \frac{2}{8} (= \frac{1}{4})$.

- Explain that to add and subtract fractions, they must have the same denominator. If they have different denominators we need to convert one or both using equivalent fractions.

- Work through some examples, in which one denominator is a multiple of the other.

◆ $\frac{1}{4} + \frac{1}{2}$: Ask: **Which fraction should we convert?** (*Change half into quarters: $\frac{2}{4} + \frac{1}{4} = \frac{3}{4}$.*)

◆ $\frac{9}{10} - \frac{1}{2}$: Ask: **Do we need to convert tenths into fifths or fifths into tenths? Which is the smaller denominator?** (*Change $\frac{2}{5}$ to tenths: $\frac{9}{10} - \frac{4}{10} = \frac{5}{10}$.*) **What is $\frac{5}{10}$ equivalent to?** ($\frac{1}{2}$)

◆ $\frac{2}{3} + \frac{9}{9} = \frac{6}{9} + \frac{5}{9} = \frac{11}{9}$: Ask: **What is $\frac{11}{9}$ as a mixed number?** ($1\frac{2}{9}$)

◆ $\frac{3}{4} - \frac{5}{12} = \frac{9}{12} - \frac{5}{12} = \frac{4}{12}$: Ask: **Can $\frac{4}{12}$ be simplified?** ($= \frac{1}{3}$)

- Show the children how to use equivalent fractions and to simplify where necessary. Use Resource 15: Fraction wall and a ruler for support to convert and simplify fractions (see Unit 22). Prompt the children to do the work. Ask: **Which fraction do we need to convert? Can we change this answer into a simpler fraction? Is the answer more or less than 1? How can you tell? What is this fraction as a mixed number?** Ensure that they work on both additions and subtractions.

- Extend to addition and subtractions of fractions where both fractions must first be converted. Use only examples up to twelfths so they can use Resource 15: Fraction wall.

◆ $\frac{3}{4} - \frac{1}{3}$: Say: **3 and 4 are both factors of 12; 12 is a multiple of 3 and of 4, so we convert to twelfths: $\frac{9}{12} - \frac{4}{12} = \frac{5}{12}$.**

◆ $\frac{1}{2} + \frac{4}{5}$: Ask: **What is a common multiple of 2 and 5?** (*10, 20…*) Say: **Convert both to tenths: 5 × 2 = 10, so multiply numerator and denominator of $\frac{1}{2}$ by 5, or use your knowledge of equivalents of $\frac{1}{2}$: $\frac{1}{2} \equiv \frac{5}{10}$. 2 × 5 = 10, so multiply numerator and denominator of $\frac{4}{5}$ by 2: $\frac{8}{10}$. Now add them together: $\frac{5}{10} + \frac{8}{10} = \frac{13}{10}$. What is $\frac{13}{10}$ as a mixed number?** ($1\frac{3}{10}$; $\frac{1}{2} + \frac{4}{5} = 1\frac{3}{10}$)

Unit 26: Multiply proper fractions and mixed numbers by whole numbers, supported by materials and diagrams

Content domain reference: 5F5

Prerequisites for learning

Multiply mentally
Understand and use equivalent fractions
Convert improper fractions to mixed numbers

Learning outcomes

Multiply a proper fraction by a whole number
Multiply a mixed number by a whole number

Key vocabulary

Proper fractions, mixed number, improper fraction, equivalent fraction

Resources

Resource 16: Fraction circles; two sets of 1–9 digit cards; coloured pencils; small squares of paper or sets of fraction cards: $\frac{1}{2}, \frac{1}{4}, \frac{3}{4}, \frac{1}{3}, \frac{2}{3}, \frac{1}{5}, \frac{2}{5}, \frac{3}{5}, \frac{4}{5}, \frac{1}{6}, \frac{5}{6}, \frac{1}{8}, \frac{3}{8}, \frac{5}{8}, \frac{7}{8}$

Background knowledge

Multiplying a fraction by a whole number is introduced by linking it with repeated addition, shown diagrammatically. Therefore children should be shown that the answer can also be worked out by multiplying **just the numerator** by the whole number. This is because the (hidden) denominator in a whole number is 1, so multiplying the denominators will result in the same denominator as the fraction denominator. Later on children will learn to multiply two fractions by multiplying the numerators to get the new numerator and multiplying the denominators to get the new denominator.

Teaching activity 1a (20 minutes)

Multiply a proper fraction by a whole number

- Ask: **What is 3 × 1?** (3); **3 × 10?** (30 or 3 tens); **3 × 100?** (300 or 3 hundreds) **So what is three lots of $\frac{1}{5}$?** (three-fifths, $\frac{3}{5}$) If appropriate, use an enlarged copy of Resource 16: Fraction circles, shading the first of the fifths circles to show that three lots of $\frac{1}{5}$ is simply $\frac{3}{5}$. Repeat with five lots of $\frac{1}{3}$, recording as $\frac{5}{3}$ and $1\frac{2}{3}$.

- Ask: **What is 2 × 4?** (8) **How many tens is 2 × 40?** (8 tens) **How many hundreds is 2 × 400?** (8 hundreds) **So how many fifths is 2 × $\frac{4}{5}$?** (8 fifths) Repeat with 3 × $\frac{2}{3}$. Ask: **How many thirds is 3 × $\frac{2}{3}$?**, shading the relevant circles on Resource 16: Fraction circles, if necessary.

- Use digit cards to generate some fraction multiplications: remove the number 1 digit card from one set and lay it on the table to be the numerator of some unit fractions. Write × on a slip of paper and put it to the right of the 1 digit card. Shuffle the other cards (one set) and turn one over to become the denominator and the next to become the multiplier, for example:

- Then work out the answer: four lots of one fifth is simply four-fifths. Use digit cards from the other set to record the calculation and the answer.

- Shuffle the 2–9 cards and repeat.

- Now remove the digit card 1 completely and select three cards to make a fraction with a numerator of 2 or more. Say: **We are going to find out how many different fraction multiplications we can make with each set of three cards.** Explain that the fraction must be a proper fraction. Digit cards 3, 4 and 9 could produce: $\frac{4}{9} \times 3, \frac{3}{9} \times 4, \frac{3}{4} \times 9$. Work out each one, recording the answers as mixed numbers.

- Encourage children to make the connection with the basic multiplication fact. Say: **Three lots of 4 make 12, so three lots of four-ninths will make 12-ninths. $\frac{12}{9}$ as a fraction is $1\frac{1}{3}$ as a mixed number.** Ask: **Which multiplications give the same answer?** ($\frac{4}{9} \times 3$ *and* $\frac{3}{9} \times 4$ *both give* $\frac{12}{9}$ *which is* $1\frac{1}{3}$) Repeat with another set of three cards.

Teaching activity 1b (15 minutes)

Multiply a proper fraction by a whole number

- Write $5 \times \frac{1}{2}$ on a whiteboard. Demonstrate how to rewrite this as a repeated addition: $\frac{1}{2} + \frac{1}{2} + \frac{1}{2} + \frac{1}{2} + \frac{1}{2}$. Using Resource 16: Fraction circles , shade half of the first , then half of the next, and so on, counting how many you have shaded: one half, two halves... five halves. Say: **Five lots of one half is simply 5 halves.** Ask: **What is this as a mixed number?** ($2\frac{1}{2}$) Repeat for $7 \times \frac{1}{3}$, $3 \times \frac{1}{4}$, $8 \times \frac{1}{5}$, supporting children to shade in the correct number of sections and recording the answer as an improper fraction and as a mixed number. Emphasise that, in the same way as seven lots of 1 is **7**, seven lots of $\frac{1}{3}$ will be **7** thirds ($\frac{7}{3}$).

- Write $2 \times \frac{3}{4}$ on a whiteboard and demonstrate, on Resource 16: Fraction circles, how to use the quarter-circles to shade in two lots of $\frac{3}{4}$. Highlight the connection with addition, to ensure that they do not think the answer is $\frac{6}{8}$. Record $2 \times \frac{3}{4} = 1\frac{2}{4} = 1\frac{1}{2}$. Repeat with $3 \times \frac{2}{5}$, $4 \times \frac{3}{10}$... Say: **Look carefully at the product of the whole number and the numerator. What do you notice?** (*It is the new numerator.*) Ask: **Has the denominator changed? Can you work out $7 \times \frac{3}{10}$ without the diagrams?** (*7 × 3 tenths: $\frac{21}{10} = 2\frac{1}{10}$*)

Teaching activity 2a (10 minutes)

Multiply a mixed number by a whole number

- Explain that to multiply mixed numbers we need to multiply both the whole number part and the fraction. Show children how to partition a mixed number into its whole number and fractional parts.

  - $3 \times 1\frac{1}{4} = 3 \times 1 + 3 \times \frac{1}{4} = 3 + \frac{3}{4} = 3\frac{3}{4}$
  - $4 \times 2\frac{3}{5} = 4 \times 2 + 4 \times \frac{3}{5} = 8 + \frac{12}{5} = 8 + 2\frac{2}{5} = 10\frac{2}{5}$

- Ask children to use partitioning to work out $5 \times 1\frac{1}{6}$, $3 \times 2\frac{3}{4}$, $5 \times 3\frac{2}{9}$.

Teaching activity 2b (15 minutes)

Multiply a mixed number by a whole number

- Remind children how to partition a mixed number into its whole number and fractional parts, writing each part on a separate slip of paper or using sets of fractions cards. Write the fraction numbers about half the size of the single digit: $1\frac{1}{4}$; 1 and $\frac{1}{4}$.

- Write $3 \times 1\frac{1}{4}$ on a whiteboard and say: **I need three lots of 1 and three lots of $\frac{1}{4}$.** Make two more sets of 1 and of $\frac{1}{4}$ on slips of paper. Move the whole numbers together and move the fractions together. Ask: **How many 1s are there?** (3) **How many quarters?** (3) **How can I record this as mixed number?** ($3\frac{3}{4}$) Repeat for another mixed number with a unit fraction, such as $4 \times 2\frac{1}{3}$. Write four sets each of 2 and $\frac{1}{3}$. Ask: **I have 2, 4, 6, 8 whole ones plus 4 thirds. What is $\frac{4}{3}$ as a mixed number?** ($1\frac{1}{3}$) Write the addition: $8 + \frac{4}{3} = 8 + 1\frac{1}{3} = 9\frac{1}{3}$.

- Extend to non-unit fractions, keeping both the whole-number part and multiplier small. For example, for $3 \times 1\frac{2}{5}$: make three sets each of 1 and of $\frac{2}{5}$. Say: **Three whole ones plus 2, 4, 6 fifths: $3 + \frac{6}{5} = 3 + 1\frac{1}{5} = 4\frac{1}{5}$. $3 \times 1\frac{2}{5} = 4\frac{1}{5}$.** Repeat with $4 \times 2\frac{3}{4}$.

Unit 27: Read and write decimal numbers as fractions

Content domain reference: 5F6a

Prerequisites for learning

Know that $10 \times 0.01 = 0.1$ and $\frac{1}{10} = \frac{10}{100}$.

Key vocabulary

Tenths, hundredths, equivalent, convert

Learning outcomes

Read and write decimal numbers as fractions of tenths and hundredths

Resources

Resource 14: Hundredths (1 per child, 2 enlarged to A3); Resource 17: Tenths number lines

Background knowledge

Decimals, fractions and percentages are different ways of showing the same values. They can sometimes be used interchangeably, but one form often seems more sensible, depending on the context. A useful piece of homework could be to look for examples of decimals, fractions and percentages in newspapers to see where and when each is used. Some common equivalents are not expressed in tenths or hundredths because they have been simplified.

Teaching activity 1a (15 minutes)

Read and write decimal numbers as fractions (tenths and hundredths)

- Remind children that our decimal number system is base 10. So 10 ones make 1 ten, 10 tens make 1 hundred, and so on. Write 0·1 on a whiteboard and read it to the children: **Zero point one. How much are ten 0·1s worth?** (*1*) Write 0·01 and read it: **Zero point zero one. How much are ten 0·01s worth?** (*0·1*) **How many 0·01s will make 1 whole one?** (*100*)

- Show an enlarged copy of Resource 15: Hundredths. Shade one column. Ask: **How many columns are there altogether?** (*10*) **What fraction of the whole square is each column?** ($\frac{1}{10}$) Explain that as ten 0·1s also make one whole, then 0·1 is equivalent to one-tenth.

- Shade the second column. Ask: **What fraction is shaded now?** (*two-tenths or one-fifth*) **How can we write this as a decimal?** (*0·2*)

- Shade the third column. Ask: **What fraction is shaded now?** (*three-tenths*) **How can we write this as a decimal?** (*0·3*)

- Repeat for four columns, identifying four-tenths, two-fifths, 0·4. Continue, encouraging children to predict the remaining fractions, without shading.

- Write all the equivalents for tenths as fractions, in tenths or fifths as appropriate, and as one-place decimals.

- Now use another copy of Resource 15: Hundredths and shade just one cell. Ask: **What fraction is shaded?** (*one hundredth, $\frac{1}{100}$*) Say: **As ten 0·01s**

make 0·1 and one hundred 0·01s make 1, we can say that this cell is also 0·01, so $0.01 \equiv \frac{1}{100}$.

- Repeat for 0·02 (2 hundredths) through to 0·09 (9 hundredths). Recap that 10 hundredths is 1 tenth.

- Shade the first three columns and five extra cells at the top of the fourth column, to show 0·35 and $\frac{35}{100}$. Repeat for other examples, recording each as a decimal and as a fraction.

- Now let the children complete Resource 15: Hundredths by shading to show 0·9, 0·8, 0·25, 0·11, 0·73, labelling each diagram as a decimal and fraction (simplified if possible). Ask: **Can we write $\frac{25}{100}$ as a simpler fraction?** ($\frac{1}{4}$)

- Check children's understanding by asking: **What is 0·67 as a fraction? What is 0·3?** ($\frac{67}{100}$, $\frac{30}{100}$ or $\frac{3}{10}$) What is $\frac{7}{10}$ as a decimal? What is $\frac{31}{100}$? (*0·7, 0·31*)

- Show the children the quick way to convert a decimal to a fraction.
  - Make one-place decimals into fractions of 10, for example, 0·4 is $\frac{4}{10}$.
  - Make two-place decimals into fractions of 100, for example, 0·43 is $\frac{43}{100}$.

  Sometimes the fraction can be simplified.

Teaching activity 1b (15 minutes)

Read and write decimal numbers as fractions of tenths and hundredths

- Distribute Resource 17: Tenths number lines and ask children to write 0 and 1 at either end of the first three number lines. They also write the word 'decimal' against the first line and the word 'fraction' against the next two.

- Point to the first 0 to 1 number line. Ask: **What number will come halfway?** (*0·5*) Say: **Remember that this is a decimal number line.** Ask: **What number is this first marker?** (*0·1*) **What comes next?** (*0·2*) Children label each marker: 0·1, 0·2 to 0·9.

- Point to the second number line. Say: **This is a fraction number line.** Ask: **What should we write halfway?** ($\frac{1}{2}$) There are ten equally spaced markers, so what fraction should be write against this first marker? ($\frac{1}{10}$) The next marker? ($\frac{2}{10}$) Children complete the number line from $\frac{1}{10}$ to $\frac{9}{10}$.

- Tell the children that decimals and fractions are different ways of showing the same values. Say: **We can see that 0·1 is equivalent to one-tenth, 0·2 is equivalent to two-tenths.** Ask them to write out the nine equivalent decimal and common fractions.

- Ask children to look at their third number line and colour every other marker, starting at the second one (0·2). Ask: **How many markers do we need to label now?** (*5*) **What fraction will each one be?** (*one-fifth*) Ask them to label them $\frac{1}{5}, \frac{2}{5}, \frac{3}{5}, \frac{4}{5}$.

- Ask: **Which decimal is $\frac{1}{5}$ equivalent to?** (*0·2*) Let them check, by using a ruler or simply by looking back at the first number line. Repeat for all the fifths and write them out as equivalent pairs: $\frac{2}{5}$ = 0·4, $\frac{3}{5}$ = 0·6, $\frac{4}{5}$ = 0·8. Ask: **Why are one-fifth and two-tenths both equivalent to 0·2?** (*because* $\frac{1}{5} \equiv \frac{2}{10}$) **Which two fractions are equivalent to 0·4?** ($\frac{4}{10}, \frac{2}{5}$) **To 0·6?** ($\frac{6}{10}, \frac{3}{5}$) **To 0·8?** ($\frac{8}{10}, \frac{4}{5}$)

- Children label the next number line from 0 to 0·1, writing 'decimal' as before. Say: **This is a 0 to 0·1 decimal number line.** Ask: **What number will come halfway?** (*0·05*) **What number is this first marker?** (*0·01*) **What comes next?** (*0·02*) Children label each marker, writing 0·01, 0·02 to 0·09.

- Children label the next number line from 0 to $\frac{1}{10}$, writing 'fraction' as before. Say: **This is a 0 to $\frac{1}{10}$ number line.** Ask: **How many hundredths is $\frac{1}{10}$?** ($\frac{10}{100} = \frac{1}{10}$) Write $\frac{10}{100}$ by the $\frac{1}{10}$ marker. Ask: **What about this first marker?** ($\frac{1}{100}$) **What comes next?** ($\frac{2}{100}$) Children label each marker, from $\frac{1}{100}$ to $\frac{9}{100}$.

- Ask: **What equivalence statement can we make?** (*0·01 = $\frac{1}{100}$, 0·02 = $\frac{2}{100}$: … 0·09 = $\frac{9}{100}$*)

- Continue, labelling the ends of the next line 0·1, $\frac{1}{10}$, $\frac{10}{100}$ and 0·2, $\frac{2}{10}$, $\frac{20}{100}$ and then marking the intermediate points to show 0·11, 0·12… 0·19 and their related fractions in hundredths.

- Check children's understanding by asking: **How do we write 37 hundredths, 93 tenths, 8 tenths as a decimal? How do we write 0·7, 0·4, 0·13, 0·99 as a fraction?** Include some decimals in which the fraction will need simplifying, such as 0·25, 0·5, 0·75 simplified to $\frac{1}{4}, \frac{1}{2}, \frac{3}{4}$ respectively.

Note: Some children may find Resource 14: Hundredths easier to use for the numbers with two decimal places, although it is very useful to label decimal number lines to reinforce their understanding of two decimal places.

- Ask: **Is there a quick way to convert a decimal to a fraction?** Make one-place decimals into fractions out of 10. Make two-place decimals into fractions out of 100. Sometimes the fraction can be simplified.

Unit 28: Recognise and use thousandths and relate them to tenths, hundredths and decimal equivalents

Content domain reference: 5F6b

Prerequisites for learning

Recognise and use tenths and hundredths and relate them to ones, tenths and decimal equivalents

Divide any number by 1000

Learning outcomes

Understand thousandths in relation to the third decimal place

Key vocabulary

Thousandths, hundredths, tenths, metric

Resources

Resource 10: Decimal place-value grids 1; calculators (optional)

Background knowledge

An understanding of the third decimal place is essential for the metric system of measure (Unit 35). Children do not see these numbers as often as two-place decimals (money), so relating them to these contexts (measures in kilometres, litres and kilograms) can help children to make sense of them.

Thousandths are another place value. The value of each decimal place is 10 times smaller than the value of the place to the left of it: 1 thousandth is one-hundredth of a tenth and one-tenth of a hundredth. It is important to read decimal places correctly: 0·365 is zero point three six five, not zero point three hundred and sixty-five. (See activity 1a below.)

Teaching activity 1a (15 minutes)

Understand thousandths in relation to the third decimal place

- Write a decimal number on a whiteboard and ask children to say it, correcting any errors. Reading three-place decimals as a three-digit number can cause confusion because the first decimal place is tenths, not hundredths, the second is hundredths, not tenths and the third is thousandths, not ones. Ask: **Where have you seen numbers with three decimal places?** (*metric measures*)

- Ask: **How would you write one-thousandth as a fraction?** ($\frac{1}{1000}$) **Two thousandths, three thousandths... nine thousandths?** ($\frac{2}{1000}$, $\frac{3}{1000}$... $\frac{9}{1000}$) Then ask: **What about ten thousandths? Can that be simplified?** ($\frac{10}{1000} = \frac{1}{100}$) Ask: **How do we write $\frac{1}{100}$ as a decimal?** (*0·01*)

- Recap that hundredths are made by dividing a single-digit number by a hundred and tenths are made by dividing a single-digit number by 10. Ask children to use a calculator (or their knowledge) to work out all single-digit numbers divided by 1000: 1 ÷ 1000 (0·001), 2 ÷ 1000 (0·002) to 9 ÷ 1000 (0·009). Say: **Dividing a single-digit number by 1000 gives the third decimal place, which is why we call them thousandths.**

- Ask children to suggest how to write the pairs of equivalent fractions and decimals for thousandths: $\frac{1}{1000} = 0·001$, $\frac{2}{1000} = 0·002$... $\frac{9}{1000} = 0·009$.

- Continue by writing $\frac{10}{1000} = \frac{1}{100}$ and establish, therefore, that $\frac{1}{100} = \frac{10}{1000} = 0·01(0)$.

- Now look at fractions of thousandths with a two-digit numerator, using division or the calculator to check: $\frac{11}{1000} = 11 \div 1000 = 0.011$, $\frac{12}{1000} = 0.012$. Try some thousandth fractions with prime-number numerators, such as $\frac{43}{1000}$, establishing that a two-digit number of thousandths can be written with digits in the hundredths and thousandths places. Show some examples with numerators that are multiples of 10, explaining that there is no need to write the final zero in the decimal fraction when the numerator is a multiple of 10. For example, 0·02 is equivalent to 20 thousandths or 2 hundredths:

$$\frac{20}{1000} = 0.02.$$

- Work through some more thousandth fractions with two-digit numerators and then extend to thousandth fractions with three-digit numerators, establishing that there will now be a digit in the tenths place. When the numerator of the fraction includes a zero in the final position, the decimal fraction will have only two decimal places. When the numerator has zeros in the final two positions, the decimal fraction will have only one decimal place. Reiterate that 100 thousandths = 1 tenth and 10 thousandths = 1 hundredth.

- Explore different numbers of thousandths: $\frac{165}{1000} = 0.165$, $\frac{250}{1000} = 0.25$ (multiple of 10), $\frac{200}{1000} = 0.2$ (multiples of 100).

Teaching activity 1b (15 minutes)

Understand thousandths in relation to the third decimal place

- Explain that, just as the digits in a whole number can go on and on, getting larger and larger in value, so the decimal places can go on and on, getting smaller and smaller in value. Say: **We are going to explore the third decimal place.**

- Give each child a copy of Resource 10: Decimal place-value grids 1. Ask them to look at the whole-number titles, then the decimal titles and ask them to suggest what the title of the untitled third decimal place column could be. Agree that it will be related to the word 'thousand', so it is 'thousandths'.

- Write a number with three decimal places and read it together with the children, ensuring that they read the decimal part correctly. Explain why this is important. Say: **Reading a three-place decimal as a three-digit number is wrong, because the first decimal place is tenths, not hundredths, the second is hundredths, not tenths and the third is thousandths, not ones.**

- Go through the number, stressing the value of each digit and writing the first two decimal places as a fraction over 10 and 100 respectively. Ask: **How will we write the third decimal place digit as a fraction?** (a fraction over 1000)

- Continue with other numbers with three non-zero decimal places, concentrating on the place value of the decimals only and writing them as decimal and fraction equivalents.

- Now ask children to write a number with a zero in the third decimal place, but still write the fraction out of 1000, for example, $0.250 = \frac{250}{1000}$. Explain that, with decimals, any zeros at the end of a number can be left out. Say: **0·250 has the same value as 0·25, so we only need to write 0·25.** Work through a few examples of this type

- Finally, work through some with zeros in the second and third places, for example, 0·200: $0.2 = \frac{200}{1000}$, saying: **0·200 has the same value as 0·2, so we only need to write 0·2.**

Unit 29: Round decimals with two decimal places to the nearest whole number and to one decimal place

Content domain reference: 5F7

Prerequisites for learning

Know the rules of rounding whole numbers

Understand money as whole numbers with two decimal places

Learning outcomes

Round decimals with two decimal places

Key vocabulary

Rounding, to the nearest, to one decimal place

Resources

Resource 17: Tenths number lines (1 per child, 1 enlarged to A3); coins (optional); dice

Background knowledge

The rules of rounding are that if the value of the digit to the right of the digit being rounded is 5 or greater, round up and if it is less than 5, round down. The same rules apply to decimals. It can be helpful to link numbers with two decimal places with money, rounding to the nearest whole pound and nearest ten pence (activity 1a). Number lines provide a visual illustration of the parameters.

Teaching activity 1a (20 minutes)

Round decimals with two decimal places

- Begin by asking children to round amounts of money to the nearest pound: £2·99, £3·15, £4·40, £2·60. Show the amounts with coins, if possible. If children are unsure, explain that it is the same as rounding to one hundred, because there are 100 pence in a pound.

- Ask: **What do you remember about rounding whole numbers?** Then explain that they have just been rounding decimal numbers to the nearest whole number, probably without realising it. Establish the rules of rounding. Ask: **What is £2·50, rounded to the nearest pound?** (£3) **Which place value do you look at to know whether it rounds up or down?** (*the ten pence*) **How much do we need, in 10ps, for the whole amount to round up?** (*50p or more*) **Which digit is this as a decimal?** (*tenths*) **What is the single-digit pence number as a decimal?** (*hundredths*)

- Write other amounts of money as numbers with two decimal places (with the pound sign) and ask children to round them to the nearest pound. Give a mixture of values, including one with 50p, such as £4·50. If children struggle, tell them to treat the numbers as if they were pounds and pence.

- Use dice to generate three single-digit numbers: use these to make a number with two decimal places, but don't include the pound sign. Round this to the nearest whole number. Repeat.

- Move on to rounding money amounts to the nearest 10p: £2·46, £3·09, £2·54, £7·81, £1·99. Ask: **What are we rounding to?** (*nearest 10*) **Which place value do you look at?** (*pence or hundredths*) **We are rounding decimals; what are we rounding to?** (*nearest tenth*)

- Use dice to generate some more sets of three digits and make numbers with two decimal places. (Again, omit the pound sign.) Ask children to round the numbers to the nearest tenth. Include at least one example that rounds to the next whole number, such as 4·99.

- Explain that when they round amounts to the nearest tenth, they must write a digit in the tenths place, even if it is a zero.

- Repeat the previous steps, this time saying: **Round these numbers to one decimal place.** Explain that is exactly the same as rounding to the nearest tenth, so there must be a digit in the first decimal place. Ask them to round 4·19 (*4·2*), 5·63 (*5·6*), 7·08 (*7·1*), 3·98 (*4·0*).

- Check children's understanding by asking them to round numbers to both the nearest whole number and to one decimal place. Include 6·95, for which the answers will be 6 then 6·0, noting the different ways of giving the answer, even though 6 and 6·0 are the same number.

Teaching activity 1b (20 minutes)

Round decimals with two decimal places

- Ask: **What do you remember about rounding whole numbers?** Establish that if the value of the digit to the right of the digit being rounded is 5 or more, they round up and if it is less than 5, they round down. Say: **To round to the nearest ten, look at the ones digit; to round to the nearest hundred, look at the tens digit.** Explain that they are going to learn how to round numbers with two decimal places. Ask: **Does anyone remember how to round a number with one decimal place to the nearest whole number?** (This was covered in Year 4.) Try a few examples to remind them: 3·2 (3), 4·6 (5), 7·5 (8). Ask: **Which digit do we need to look at?** (*tenths*) Use Resource 17: Tenths number lines to demonstrate, as below, if necessary.

- Write the number 3·25 on a whiteboard or paper. Show children an enlarged copy of Resource 17: Tenths number lines and say: **3·25 is on this number line somewhere and there is a whole number at each end. Which number will be at this end? And this end?** (*3 and 4*) **Where, approximately, on this number line will 3·25 be?** (*halfway between the second and third marker points*) **Is this nearer to 3 or to 4?** (*3 as 3·25, to the nearest whole number, is 3. 3·25 rounds down to 3.*)

- Repeat, choosing a variety of numbers, or using dice to generate them. Write the two possibilities at the ends of the number lines and estimate the position of the number on that line to illustrate whether it will round up or down. Refer each time to the tenths, as this is the significant digit when rounding to a whole number. You can also make the connection with rounding to the nearest 100: tenths worth 50 or more round up. For example, for 4·82, ask: **Which numbers need to go on either end of the line?** (*4 and 5*) **Where is 4·5 on the line? Is 4·82 more or less than 4·5?** (*more*) **Which digit is the important one here?** (*The tenths, the 8; 8 is more than 5, 82 tenths is more than 50 tenths, so 4·82 rounds up to 5.*)

- Children use Resource 17: Tenths number lines to try some on their own. After an initial few using the number lines, encourage children to try some without, using the connection with rounding to the nearest 100.

- Now explain that they are going to round to the nearest tenth, which is also called rounding to one decimal place. Consider 4·82 again. Use the next number line and ask: **Which numbers need to go on either end of the line so we can round to one decimal place?** (*4·8 and 4·9*) **What number is exactly halfway?** (*4·85*) **Is 4·82 more or less than 4·85?** (*Less, so 4·82 rounds down to 4·8; 2 is less than 5 so the tenths digit, 8, stays the same.*) Repeat with 4·87, using the same line. (*4·87 > 4·85, 7 > 5, so 4·87 rounds up to 4·9.*) Make the connection with rounding to the nearest 10: 82 tenths rounds down to 80 tenths; 87 tenths rounds up to 90 tenths.

- Repeat with more examples, using the number lines as support initially, then encourage children to round by using the hundredths digit and the connection with rounding to 10.

Unit 30: Read, write, order and compare numbers with up to three decimal places

Content domain reference: 5F8

Prerequisites for learning

Recognise and use thousandths

Relate thousandths to tenths, hundredths and fraction equivalents

Learning outcomes

Order and compare numbers with up to three decimal places

Key vocabulary

Tenth, hundredth, thousandth, compare, order, decimal place

Resources

Resource 10: Decimal place-value grids 1; digit cards 0–9

Background knowledge

Children often think that the more decimal places, the larger the number, as this is what they have seen with whole numbers. However, the further to the right the digits are, the smaller they are in value. We compare decimal numbers by comparing the digits in each place, in the same way that we compare whole numbers. If the digits before the decimal point are the same, then the first digits to compare are the tenths (activity 1a). Making the decimal parts all the same size by writing zeros in the place holders is a useful visual strategy (activity 1b).

Teaching activity 1a (15 minutes)

Order and compare numbers with up to three decimal places

- Show the children Resource 10: Decimal place-value grids 1 and ask: **What will be written in this last column?** (*thousandths*) Ask a child to suggest a tens and a ones number that has three decimal places, such as 34.286 (*thirty-four point two eight six*). Ensure that they say the decimal part correctly. Ask a different child to write it in the first grid. Recap on the value of each digit, in this example, 3 tens, 4 ones, 2 tenths, 8 hundredths, 6 thousandths.

- Ask another child to suggest a number with the same tens and ones digits but with the decimal digits rearranged, for example, 34·682. Write this in the next line of the same grid. Ask children to say this number and give the value of each digit. Ask: **Which is larger, 34·682 or 34·286?** (*34·682*) **Which digit do we use to work this out?** (*Tenths: because 34·682 has six tenths and 34·286 only has two tenths, 34·682 is larger.*) Say: **We do not need to consider the hundredths and thousandths in this example.**

- Continue by asking a child to write 34·73 in the next line of the same grid. Ask: **Which is larger, 34·682 or 34·73?** (*34·73*) **How many tenths does 34·682 have?** (*6*) **How many tenths does 34·7 have?** (*7*) **So which number is the larger?** (*34·73*) Some children will find it easier to compare decimals that are the same length, so show them how to write in any zero place holders: 34·**730**. Emphasise that 34·730 is exactly the same number as 34·73. It is now straightforward to compare them in the same way as whole numbers, 730 > 682. Ask children to use the next grid to compare 12·75, 12·749 and 12·8, ordering them from smallest to largest.

- Ask children to use the next grid to compare 26·487, 26·49, 26·4, ordering them from smallest to largest. Ask: **Which digit do we need to look at?** (*hundredths*)

- Discuss athletics events involving time and length, where scores are recorded using decimals. Ask: **Which competitor would win an event involving length – the one with the longest length or the shortest length? Which competitor would win an event involving time (running): the one with the greatest (longest) time or the least (shortest) time?**

Teaching activity 1b (15 minutes)

Order and compare numbers with up to three decimal places

- Use the digit cards to generate a tens and a ones number that has three decimal places, for example, 34·286. Say the number together, ensuring that children say the decimal part correctly: **thirty-four point two eight six**. Recap the value of each digit.

- Ask another child to rearrange the digits of the decimal part of the number, for example, 34·682. Ask them to say it and give the value of each digit.

- Write 34·286 and 34·682 next to each other, with a gap between them. Ask: **Which symbol, > or <, do we need to write here to compare the two numbers? Which digit do we need to compare? The tens and ones are the same, but are the tenths the same? Which number has more tenths?** (*34·682*) Ask a child to draw the correct symbol between the two numbers. (<)

- Ask: **Is there a different strategy we could use?** (*Use our knowledge of three-digit numbers, for example, 286 is less than 682.*)

- Repeat, but keep the tens, ones **and** tenths digits the same. Comparing 34·286 and 34·268, ask: **Which digit do we need to compare now?** (*hundredths*) Write 34·286 and 34·268 next to each other with a gap between them, as before. Ask: **Which number is larger this time?** (*34·286 has 8 hundredths, 34·268 has 6 hundredths, so 34·286 is larger than 34·268.*) Establish that 286 > 268. Ask a child to write the correct symbol.

- Generate a third number with the same tens and ones digits, but this time remove one of the decimal digits and rearrange the remaining two decimal digits, for example, 34·62, to compare it with both 34·286 and 34·682. Explain: **It can be helpful to write zero placeholders when comparing decimals of different lengths so that we can still use our knowledge of three-digit numbers.** Demonstrate, writing 34·62**0**. Say: **620 is less than 682 but more than 286, so 34·620 < 34·682; 34·620 > 34·286.** Verify this by comparing the tenths and hundredths digits. Say: **34·286 has 2 tenths, 34·682 and 34·62 both have 6 tenths, so 34·286 is the smallest. 34·682 has 8 hundredths, 34·62 has 2 hundredths, so 34·682 is the largest. Their order, smallest to largest, is 34·286, 34·62, 34·682.**

- Add a fourth number to the group, with the same tens and ones digit, but only one decimal digit (tenths), for example, 34·8. Order 34·62, 34·8, 34·286, 34·682 from **largest to smallest,** rewriting 34·8 as 34·8**00** and 34·62 as 34·62**0** as before.

- Repeat, using a different set of digit cards, but comparing pairs and then sets of numbers with the same tens and ones, but different decimals of different lengths.

- Finally, discuss athletics events where scores are recorded using decimals. Ask: **Which competitor would win an event involving length – the one with the longest length or the shortest length? Which competitor would win an event involving time (running): the one with the greatest (longest) time or the least (shortest) time?**

Unit 31: Solve problems involving number up to three decimal places

Content domain reference: 5F10

Prerequisites for learning

Add and subtract numbers by the column method

Add and subtract numbers mentally

Understand and use numbers with up to three decimal places (d.p.)

Know or work out number pairs to 100

Learning outcomes

Add and subtract numbers with up to three decimal places mentally

Add and subtract numbers with up to three decimal places by a written method

Work out decimal complements to 1 and 10

Key vocabulary

Complement to 1 and 10; place holder; halfway; between

Resources

Resource 6: 100 square; Resource 10: Decimal place-value grids 1; Resource 11: Decimal place-value grids 2; Resource 14: Hundredths; sets of digit cards 0–9; decimal point cards; 1 cm squared paper; two pieces of string of different lengths > 1 m; counters in two different colours

Background knowledge

The activities in this section focus on the skills needed to add and subtract decimals, both mentally and by written methods. Children need to be able to use these important skills to solve problems, so provide plenty of practice. The primary focus of the activities here is on the skills, although the activities include more problem solving.

The methods for adding and subtracting decimal numbers are the same as for whole numbers. Children often struggle when the numbers do not have the same number of decimal places. It is essential that they line up the decimal points so the digits with the same place values are aligned.

Teaching activity 1a (15 minutes)

Add and subtract numbers with up to three decimal places mentally

- Write $6·1 + 0·03 + 0·004$ on a whiteboard. Ask: **How can we work this out mentally?** Point out that this a 'place-value addition'. The answer is simply $6·134$. Give children some more, similar examples.

- Write $5·4 + 0·3$ and say: **We only need to add the tenths here: 4 tenths + 3 tenths is 7 tenths, so the answer is 5·7.** Continue with $7·2 + 4·6$ (*11·8*). Give children some more examples in which the sum of the tenths digits is less than 10.

- Write $3·16 + 2·53$, showing how each place value can simply be added mentally, giving the answer $5·69$. Children try $4·52 + 3·35$ (*7·87*), $7·22 + 5·74$ (*12·94*).

- Write $2·9 + 4·5$. Say: **We know that 29 + 45 = 74, so 2·9 + 4·5 = 7·4.** Children try $7·8 + 5·7$ (*13·5*), then $3·45 + 7·68$ (*11·13*), making the link with $345 + 768$ (*1113*).

- Repeat the process for subtraction: $5·4 - 0·3$ (*5·1*); $7·6 - 0·4$ (*7·2*); $7·85 - 4·13$ (*3·72*); $8·88 - 5·55$ (*3·33*).

- Show how $6·2 - 4·6$ is linked to $62 - 46 = 16$, so $6·2 - 4·6 = 1·6$. Children try similar examples in which the tenths digit in the subtrahend (number being subtracted) is smaller than the tenths digit in the minuend (the number being subtracted from).

- Where appropriate, children try $3·67 - 2·85$ making the link with $367 - 285$.

- Ask: **What is the number halfway between 7·2 and 3·8?** (*5·5*) Illustrate this on a number line as on the pupil page.

| 3.8 | ? | 7.2 |

- Explain this strategy: **Add the numbers together, then halve the answer: 7·2 + 3·8 = 11. Half of 11 is 5·5; 5·5 is half-way between 3·8 and 7·2.** Check by finding the difference between 5·5 and 3·8 (*1·7*) and between 5·5 and 7·2 (*1·7*). Another strategy would be to halve both numbers and add: $1·9 + 3·6 = 5·5$

Teaching activity 1b (15 minutes)

Add and subtract numbers with up to three decimal places mentally

- Write 6·1 + 0·03 + 0·004 on a whiteboard. Distribute Resource 10: Decimal place-value grids 1, writing thousandths in the last column. Ask: **Where do I write 6.1 in this grid?** Write it in. Ask: **Where do I write the 0·03?** (*in the second decimal place*) Write it in, pointing out that you write 3, not 0·03. Ask: **Where do I write the 0·004?** (*in the third decimal place*) Write it in, pointing out that you write 4, not 0·004. Ask: **What number have we got?** (*6·134*) Say: **These are just place-value additions – we do not need to add any digits to get the answer.** Repeat with similar examples, using the grid for the first few then asking children to work the answers out mentally.

- Write 5·4 + 0·3. Ask: **How can we add these numbers?** Make the link with whole-number addition of 54 and 3 (*57*). Say: **So 5·4 + 0·3 = 5·7.** Practise with similar examples, then try more difficult additions such as 7·2 + 4·6 (*11·8*), 2·9 + 4·5 (*7·4*) Continue using the grids for illustration and support until children can add these numbers mentally.

- Repeat with subtractions, using the grid for the first few of each type until they are successful without the grid.

- Ask: **What is the number half way between 7·2 and 3·8?**

- Explain this strategy: **Add the numbers together, then halve the answer: 7·2 + 3·8 = 11.** Half of 11 is 5·5. 5·5 is half-way between 3·8 and 7·2. Check by finding the difference between 5.5 and 3·8 (*1·7*) and between 5·5 and 7·2 (*1·7*). Another strategy would be to halve both numbers and add: 1·9 + 3·6 = 5·5.

Teaching activity 2a (20 minutes)

Add and subtract numbers with up to three decimal places by a written method

- Say: **We can add and subtract numbers with decimal places in the same way as whole numbers.** Use digit cards, with decimal point cards, to make two numbers (< 10) each with three decimal places (tenths, hundredths, thousandths). Display them side by side. Ask a child to move the cards, laying them out in the column method used in Unit 7. Say: **Each place value needs to be aligned carefully.** Ask: **Are the decimal points lined up? Let's put the decimal point in the answer line before we forget. Where should it go?** (*directly under the other two decimal points*) **Now we can add the digits in the usual way.** Children copy the sum onto their whiteboards and work out the answer. Check that they have remembered to carry over, where necessary, and added in those carried numbers. Repeat with more examples until children are confident.

- Use the same method to make two more numbers less than 10. This time, make one with two decimal places and the other with three. Let a child lay them out as a column addition. Then indicate the third decimal place in the first number, saying: **There is a gap here. What can we do?** Continue with: **We can use a zero place holder to show there are no thousandths in this number.** Let children copy and complete the sum, as before. Repeat the process with two more numbers less than 10. This time, make one with only one decimal place and the other with three.

- Work through the same process for subtraction. Remind children to put the larger of the two generated numbers first. Stress the importance of lining up the decimal points, as this will ensure that digits with the same place value are lined up correctly. Start by using two numbers with the same number of decimal places, revising the exchange method when the digit being subtracted is larger, as in Unit 7.

- Repeat with two numbers with different numbers of decimal places. Work through one or two examples in which the number being subtracted has fewer decimal places, writing zero placeholders in the relevant columns. Then demonstrate subtracting a number from a larger number with fewer decimal digits. Start by subtracting a number with tenths, hundredths and thousandths from a number with only tenths and hundredths. Say: **The zeros are very important here.** A common error is for children to write in the digits from the smaller number. After writing in the zero, children will need to exchange, to complete the subtraction in the hundredths column. Then work through some examples subtracting a number with only one decimal digit in the tenths columns from a number with tenths, hundredths and thousandths. Here, exchanges over two columns are required, similar to 400 – 175 in whole numbers.

- Show two pieces of string, saying that one measures 1·25 m and the other 1·4 m. Lay them end to end. Ask: **What calculation will tell us the total length of the string?** (*1·25 + 1·4*) Lay the longer one alongside the shorter one, so it extends beyond the end. Ask: **What calculation will tell us the difference in lengths?** (*1·4 – 1·25*) Work out both answers. Ask children to think of a different word problem that would involve adding or subtracting decimals of different lengths.

Teaching activity 2b (20 minutes)

Add and subtract numbers with up to three decimal places by a written method

- Use digit cards to generate two numbers (< 10) with three decimal places, or ask for children's suggestions. Copy the two numbers into the grid on Resource 11: Decimal place-value grids 2, larger number first, writing thousandths in the last column. Explain that numbers with decimal places can be added and subtracted in the same way as whole numbers.

- Draw a line under the two numbers for the answer line (third line of grid). Highlight that the decimal points, including the one in the answer line, are aligned. Add each column, as for whole numbers, writing any carried numbers in the fourth line.

- Give children another example to do on squared paper. Stress the importance of making sure the decimal points are lined up and that they write the decimal point in the answer line before they start adding.

- Extend the activity (3 d.p. + 2 d.p; 1 d.p. + 3 d.p.), writing in any zero place holders. Complete the first of each type of question on Resource 11: Decimal place-value grids 2, then let children work on squared paper.

- Move on to subtraction of numbers with three decimal places, revising how to exchange (Unit 7) then extend activity (3 d.p. – 1 d.p.), writing zeros in the place holders in the lower number.

- Now demonstrate mixed examples (2 d.p. – 3 d.p. and 1 d.p. – 3 d.p.), showing how to use zero place holders. Make the link with whole-number examples: 2.1 – 1.845 with 2100 – 1845.

  After each example, give the children a similar problem to complete themselves.

- Show the two pieces of string, saying that one measures 1·25 m and the other 1·4 m. Lay them end to end. Ask: **What calculation will tell us the total length of the string?** (*1·25 + 1·4*) Lay the shorter one under the longer. Ask: **What calculation will tell us the difference in lengths?** (*1·4 – 1·25*) Work out both answers. Ask children to think of a different word problem that would involve adding or subtracting decimals of different lengths.

Teaching activity 3a (15 minutes)

Work out decimal complements to 1 and 10

- Say a number and ask children to write its complement (pair) to 100, for example, you say 64, they write 36. Repeat several times. Note where children's answers are ten too many, reminding them that the tens must only add to 9, as the extra ten comes from adding the ones.

- Say: **We can use complements to 100 to find pairs (complements) of decimal numbers that add to 10 or to 1.** Show them that 64 + 36 = 100, so 6·4 + 3·6 = 10 and 0·64 + 0·36 = 1. Then consider 40 + 60 = 100: the corresponding complement to 10 is 4 + 6 = 10, which we can write as 4·0 + 6·0 = 10·0. Then we can see that 0·4 + 0·6 = 1.

- Give children, in pairs, Resource 6: 100 square and some counters. Ask one child to choose a number by putting a counter on it. The other child places their counter on its pair to 100. Then they both write the corresponding bond to 10 for one-place decimals: for example, 6·4 + 3·6 = 10. They repeat, swapping roles. They do this several more times, recording the complement each time.

- Repeat, but this time recording the complements to 1 of two-place decimals: 0.64 and 0·36 = 1. Repeat several more times.
- Write 4·7 + ☐ = 10, 0·24 + ☐ = 1, 10 – ☐ = 6·8 and 1 – ☐ = 0·12 on a whiteboard, asking children to find the missing numbers.
- Finally, ask children to see if they can work out some pairs of three-place decimals to add to 1, 10 or 100, using the fact that 174 + 826 = 1000: 0·174 + 0·826 = 1, 1·74 + 8·26 = 10, 17·4 + 82·6 = 100. Repeat by asking children to supply a pair to 1000, establishing that both the hundreds and tens will have a total of **9**, not 10.

Teaching activity 3b (15 minutes)

Work out decimal complements to 1 and 10

- Say a number and ask children to write its complement (pair) to 100, for example, you say 56, they write 44. Repeat. Note any answers that are ten too many. Say: **The tens must only add to 9, as the extra 10 comes from adding the ones.**
- On Resource 14: Hundredths, use two colours to shade the grid to illustrate a complement to 100, such as 64 + 36. Say: **We can use this to find decimal complements to 1 and to 10. Each square is one hundredth so this shows 64 hundredths or 0·64 and 36 hundredths 0·36: 0·64 + 0·36 = 1.** Children use their grids to shade in two blocks of different colours, writing the pair of decimals as a sum to 1.

- Remind children of the link with whole-number pairs to 100. Ask: **Can you work out the missing number in 0·24 + ☐ = 1?** (0·76) **And 1 – ☐ = 0·12?** (0·88)
- Explain that if the hundred grid represented **ten**, then each cell would be a tenth. Shade to show 4·5 + 5·5 = 10 (45 tenths = 4·5, 55 tenths = 5·5). Again, stress the link with whole-number pairs to 100 (45 + 55 = 100). Ask: **Can you work out the missing numbers in 4·7 + ☐ = 10?** (5·3) **And 10 – ☐ = 6·8?** (3·2)
- Say: **Two decimal numbers add together to equal 1. One of the numbers is 0·17. what is the other number?** (0·83)
- Finally, ask children to work out some pairs of three-place decimals to total 1, 10 or 100, using the fact that 174 + 826 = 1000: 0·174 + 0·826 = 1; 1·74 + 8·26 = 10; 17·4 + 82·6 = 100. Repeat by asking children to supply a pair to 1000, establishing that both the hundreds and tens places will have a total of **9**, not 10.

Unit 32: Recognise the per cent symbol (%) and understand that per cent relates to 'number of parts per hundred'; write percentages as a fraction with denominator 100, and as a decimal fraction

Content domain reference: 5F11

Prerequisites for learning

Understand fractions and decimals in tenths and hundredths

Learning outcomes

To understand percentage as parts of a hundred

Write a percentage as a fraction and as a decimal

Key vocabulary

Percentage, per cent, %

Resources

Resource 14: Hundredths

Background knowledge

Per cent means per hundred. Cent means 100, as in centimetre. 100 cents = $1. Decimals, fractions and percentages are different ways of showing the same values, so each percentage has a decimal and a fraction equivalents: 50% = $\frac{1}{2}$ because 50% is half of 100% and has been simplified from $\frac{50}{100}$. Therefore, 50% = 0.5.

Teaching activity 1a (10 minutes)

Understand percentage as parts of a hundred

- Explain what a percentage is (see Background knowledge) and how to write the symbol %. As an example, say: **If there are 100 questions in a test and someone gets 50 right, we say that they scored 50%; 75 correct is a score of 75 per cent. A score of 100% means they answered all 100 questions correctly.**

- Show children Resource 14: Hundredths, reminding them that this grid has 100 small squares, 10 rows of 10. Say: **This whole grid is 100%, so what is each square worth?** *(1%)* **What is one row or column worth?** *(10%)* Ask children to show 1%, 10%, 15%, 20%, 25% and 62% on their grids, each time writing the percentage underneath.

- Colour a set of grids yourself and ask children to identify the percentage shown.

Teaching activity 1b (10 minutes)

Understand percentage as parts of a hundred

- Remind children what a percentage is, and how to write the symbol %. Give an example: **If there are 100 questions in a test and a pupil gets 50 right, we say that they scored 50%. seventy-five correct is a score of 75 per cent. A score of 100% means they answered all 100 questions correctly.**

- Say: **100 per cent means all of something.** Explore together where percentages are used in real life. Ask: **If 100% of all the children in a school are boys, what type of school is it?** *(an all-boys school)* **Max got 90% in a test. What percentage did he get wrong?** *(100 – 90 = 10%)* **In a class, 85% of children are right-handed. What percentage are left-handed?** *(15%)* **The weather girl says there is 25% chance of rain. What is the chance that it will not rain?** *(75%)*

Teaching activity 2a (15 minutes)

Write a percentage as a fraction and as a decimal

- Show children Resource 14: Hundredths, asking: **What is each square as a fraction?** $(\frac{1}{100})$ **What is each row as a fraction?** $(\frac{1}{10})$ **What is each square as a decimal?** *(0·01)* **What about each column?** *(0·1)*

- Now ask: **What is each square as a percentage?** *(1%)* **What is each row or column as a percentage?** *(10%)* Say: **Because 1 square is $\frac{1}{100}$ and 0·01 and 1%, we can say that $\frac{1}{100}$, 0·01 and 1% are equivalent. The columns show us that $\frac{1}{10}$, 0·1 and 10% are also equivalent.**

- Explain that decimals and fractions can also be written as percentages. If the children have completed activity 1a, showing 1%, 10%, 15%, 20%, 25% and 62% on their grids, work through 1%, 10% and 15% to show how to write each of these as a fraction out of 100 (simplified for tenths only) and as a decimal (1 out of a hundred, 0·01; 10 out of a hundred = 1 out of ten, 0·1; 15 out of a hundred = 0·15). Then ask children to complete the rest. Otherwise, children should shade their grids to show 1%, 10%, 15%, 20%, 25% and write the percentage, decimal and fraction underneath in each case.

- Now encourage children to write other percentages as fractions of 100 or 10 and as decimals, providing Resource 14: Hundredths only if necessary. Ask them to show: 30%, 54%, 80%, 75%, 99%, 5%.

- Ask: **How do I change a percentage into a fraction?** *(Use the digits in the percentage as the numerator of a fraction with a denominator of 100.)*

- Ask: **How do I change a percentage into a decimal?** *(Divide the percentage by 100.)* Include examples with single-digit percentages, such as 5% = 0·05.

- Work through some examples of converting fractions (tenths and hundredths) and decimals to percentages.

Teaching activity 2b (15 minutes)

Write a percentage as a fraction and as a decimal

- Explain that decimals, fractions and percentages are just different ways of showing the same values, so every percentage has a decimal and a fraction equivalent. Say: **Per cent means 'out of 100', so to write a percentage as a fraction, you write the percentage as a fraction out of 100: 21% is $\frac{21}{100}$, 50% is $\frac{50}{100}$ (= $\frac{1}{2}$).**

- Convert some more percentages into fractions of 100, only simplifying with tenths or known equivalents. Let children try 45%, 80%, 23%, 67%, 60%, 6%, 2%.

- Say: **To convert a percentage to a decimal, divide by 100.** Give some examples.

$$21\%: 21 \div 100 = 0·21$$
$$50\%: \div 100 = 0·50 (= 0·5)$$
$$5\%: 5 \div 100 = 0·05$$

Commenting on the final example (5% = 0·05), say: **Note the zero. Single-digit percentages are only hundredths, so we need this zero.**

- Convert some more percentages into decimals by dividing by 100. Ask: **What do you notice about the digits?** *(They are the same, but the percentage becomes the first two decimal places.)* Say: **Usually, a percentage produces a decimal with two decimal places, but a single-digit percentage, such as 5% = 0.05, produces a decimal with zero tenths.** Then say: **You can write a percentage that is a multiple of ten, such as 50%, with just one decimal place: 50% = 0.5.** Ask children to write some more percentages as decimals: 45%, 80%, 23%, 67%, 60%, 6%, 2% as above.

- Ask children, in pairs, to explain to each other how to convert a percentage into a decimal or into a fraction.

Unit 33: Solve problems which require knowing percentage and decimal equivalents of $\frac{1}{2}$, $\frac{1}{4}$, $\frac{1}{5}$, $\frac{2}{5}$, $\frac{4}{5}$ and those fractions with a denominator of a multiple of 10 or 25

Content domain reference: 5F12

Prerequisites for learning

Express a percentage as a fraction and as a decimal

Understand equivalence in fractions

Learning outcomes

Convert between fractions and percentages to solve word problems

Key vocabulary

Percentage, equivalent fractions

Resources

Resource 14: Hundredths; 1 cm squared paper

Background knowledge

Although percentages, decimal and fractions are different ways of displaying the same information, they cannot all be used interchangeably. Some contexts are best described as a percentage, some as a fraction and some as a decimal. Try to find examples of each in the media to illustrate some of these contexts. As KS2 children do not use calculators, they will need to use fractions that can easily be converted to tenths or hundredths.

Teaching activity 1a (20 minutes)

Convert between fractions and percentages to solve word problems

- Ask children to draw two horizontal lines, each 10 cm long, on squared paper, one below the other, with a space of one or two squares between them. Then ask them to write 0 and 1 at either end of the first line and 0 and 100% at either end of the second line.

- Recap that the number 1 is a whole one and that 100% also means the whole. Ask children to mark halfway points on both lines. For the first line, say: **The top line is a fraction line so what fraction do we need to write at this halfway marker?** ($\frac{1}{2}$, *half*) For the second line say: **This bottom line is the percentage line, so what percentage will this halfway marker be?** (*50%*) Tell the children that 50% is equivalent to one half.

- Next, ask the children to mark the quarters. The first ($\frac{1}{4}$) should be between the second and third squares. The second ($\frac{3}{4}$) should be between the seventh and eighth squares. Ask them to write the fraction and percentage against each mark ($\frac{1}{4}$, 25%; $\frac{3}{4}$, 75%). **Ask: What percentage is $\frac{1}{4}$ equivalent to?** (*25%*) **What fraction is 75% equivalent to?** ($\frac{3}{4}$) Remind children that these are also equivalent to 0·25 and 0·75.

- Write the three sets of equivalents under the lines: $\frac{1}{2}$ is the same as 50% is the same as 0·5; $\frac{1}{4}$ is the same as 25% is the same as 0·25; $\frac{3}{4}$ is the same as 75% ≡ 0·75.

- Ask children to draw two more horizontal lines 10 cm long and label them as before: 0 to 1 and 0 to 100%. Mark each square to show tenths. **Ask: What do we write against this first marker on the fraction line?** (*one-tenth*) Children label the first line in tenths as fractions ($\frac{1}{10}$, $\frac{2}{10}$, up to $\frac{9}{10}$). Ask the children to work out what each marker needs to be on the percentage line (10%, 20% up to 90%). Children label the second line in multiples of 10%.

- Look at the one-tenth and 10% and say: **They are in the same place on the number lines, they are equivalent. What would it be if it was a decimal number line?** (*0.1, Unit 27*) Everyone writes the set of three equivalents as before and continues with all the others: $\frac{1}{10} \equiv 10\% \equiv 0{\cdot}1$, $\frac{2}{10} \equiv \ldots$

- It is worth pointing out that as $\frac{2}{10} \equiv \frac{1}{5}$ we can also write: $\frac{1}{5} \equiv 20\% \equiv 0{\cdot}2$. Ask children to write the other sets of equivalents for fifths.

- Show how to convert scores out of 25 and 50 to an equivalent fraction in hundredths, using doubling. For 45 out of 50, say: **50 is half of a hundred, so doubling both numbers gives 90 out of 100, which is 90%; 21 out of 25 = 42 out of 50 = 84 out of 100, so 84%.**

- Finally, link this to some word problems: **Jemma got nine out of ten in a maths test. What is this a percentage?** (*90%*) **Becci got 35 out of 50. What is her score as percentage?** (*70%*) **To make 50 as a percentage, you multiply it by 2, to make 100, then you multiply 35 by 2. Jess got 23 out of 25. What is her score as a percentage?** (*92%*)

Teaching activity 1b (20 minutes)

Convert between fractions and percentages to solve word problems

- Use Resource 14: Hundredths to work out the percentage equivalents for a quarter, half and three-quarters, by shading the fraction on the grids and counting the squares, or by using equivalent fractions to show that $\frac{1}{2} \equiv \frac{50}{100} \equiv 50\%$, $\frac{1}{4} \equiv \frac{25}{100} \equiv 25\%$, $\frac{3}{4} \equiv \frac{75}{100} \equiv 75\%$.

- Repeat for tenths, shading the columns to show $\frac{1}{10} \equiv \frac{10}{100} \equiv 10\%$, $\frac{2}{10} \equiv \frac{20}{100} \equiv 20\%$.

- Consider some word problems involving 50, 25 and 20, for example, asking: **Rich scored 42 out of 50 in a maths test. What is his score as a percentage?** Explain: **42 out of 50 can be written as the fraction $\frac{42}{50}$ and by doubling both parts of the fraction we get an equivalent fraction $\frac{84}{100}$. What is $\frac{84}{100}$ as a percentage?** (*84%*) Repeat for other scores out of 50.

- Ask: **How do you write 21 out of 25 as a fraction?** ($\frac{21}{25}$) **How could we make an equivalent fraction in hundredths?** (*Doubling and doubling again or multiplying by 4.*) Write down the steps: $\frac{21}{25} (\equiv \frac{42}{50}) \equiv \frac{84}{100} \equiv 84$. Repeat for other scores out of 25.

- Ask: **How do you write 16 out of 20 as a fraction? How could we make an equivalent fraction in hundredths?** (*5 × 20 = 100, so multiply both parts by 5 or halve and multiply by 10: $\frac{16}{20} \equiv \frac{80}{100} \equiv 80\%$.*) Repeat for other scores out of 20.

Number – fractions (including decimals and percentages)

Unit 34: Convert between different units of metric measure (for example, kilometre and metre; centimetre and metre; centimetre and millimetre; gram and kilogram; litre and millilitre)

Content domain reference: 5M5

Prerequisites for learning

Multiply and divide any number by 10, 100 and 1000

Be familiar with the metric measures of length, mass and capacity and their contractions

Learning outcomes

Convert between millimetres, centimetres and metres

Convert between g/kg, m/km and ml/litres

Key vocabulary

Convert, metric, length, distance, mass, capacity, kilometre, kilogram, litre, gram, metre, centimetre, millimetre, millilitre

Resources

Resource 18: Conversion tables; cm/mm rulers; metre wheels; metre sticks; tape measures (longer than 1 m), kg scales; measuring containers: jugs, plastic milk containers, packets, tins, jars (optional)

Background knowledge

Children will have experience of working with metric measures but can often not remember the conversion factor, even though there is only a choice of 10, 100 or 1000. Having practical experience of what each unit looks like is very important. For some children, knowing the roots of the prefixes kilo- (1000), milli-(1000), centi-(100) may help. For milli**metre**, the 1000 is linked to metre, not centimetre: 1000 mm = 1 m; for milli**litre** it is linked to litre: 1000 ml = litre. If the conversion involves **cm**, then the conversion factor is **not** 1000. For all the others used in KS2 the conversion factor **is** 1000.

Teaching activity 1a (15 minutes)

To convert between millimetres, centimetres and metres

• Ask children to look carefully at a 15 cm or 30 cm ruler. Ask: **What does a ruler measure?** (*length*) **Which units of length can you see?** (*centimetres and millimetres*) **Which is smaller, a millimetre (mm) or a centimetre (cm)?** (*mm*) **How many millimetres are there in each centimetre?** (*10*) **How many millimetres are there in 2 cm? In 5 cm? In 11 cm?** Record as 1 cm = 10 mm, 2 cm = 20 mm… Ask: **How could we use multiplication to work out how many millimetres there are in 20 cm without looking at a ruler?** (*Multiply the number of cm by 10.*) **What length, in millimetres, is equivalent to 20 cm?** (*200 mm*) **And 45 cm?** (*450 mm*)

• Repeat for converting millimetres to centimetres, using the ruler at first. Ask: **How can we convert 30 mm to centimetres? What length, in centimetres, is equivalent to 50 mm?** (*5 cm*) **To 100 mm?** (*10 cm*) **To 120 mm?** (*12 cm*)

• Look at the ruler again, considering the markers between whole centimetres. Ask: **How many centimetres is the halfway point between 2 cm and 3 cm?** (*two and a half*) **How do we write this as a decimal?** (*2·5*) **How many millimetres are there in 2.5cm?** (*25*) Record as 2·5 cm = 25 mm. Choose some other measurements 'in between' whole centimetres, writing them with one decimal place (d.p.) and converting them to millimetres with – and then without – the aid of the ruler.

• Repeat by converting two-digit millimetre measurements into centimetres by dividing by 10 and recording, for example, 34 mm = 3·4 cm. Establish that 1 mm is very small (fingernail thickness) and 1 cm is approximately the width of a finger, so it is still quite small.

• Show the children a metre stick and tell them that this is a metre. Hold it up against a child and ask: **Is .... taller or shorter than 1 metre?** Ask other questions: **Is the height of the door more or less than 1 m? Is it more or less than 2 m?** Use two metre sticks to find out. Ask children to look at the metre sticks to discover or recap how many centimetres there are in 1 m: 100 cm = 1 m.

- Use a tape measure to convert some metre measurements to centimetres, including metres as decimals, for example: 1·3 m = 130 cm. Then convert lengths in centimetres to metres, for example: 300 cm = 3 m, 145 cm = 1.45 m, establishing that the number to multiply or divide by is 100. Use the stick to show that 0·5 m = 50 cm, 0·3 m = 30 cm, 90 cm = 0·9 m…

Teaching activity 1b (15 minutes)

To convert between millimetres, centimetres and metres

- Show children the ruler, to show that 1 mm is very small, 1 cm is larger and 1 m is much larger. It helps to link each length with a real object, to have an image of what each looks like; for example, 1 mm: top edge of a fingernail; 1 cm: width of a finger; 1 m: height of child.

- Explain that it is easy to convert between metric units because we only need to decide if the conversion factor is 10, 100 or 1000. Ask: **How many millimetres are there in each centimetre, 10, 100 or 1000?** (*10*) Say: **This means we can multiply or divide by 10 to change between millimetres and centimetres.**

- Sketch a function machine in the shape of a ruler and write 'cm' on the left and 'mm' on the right, with × 10 over the ruler and ÷ 10 underneath.

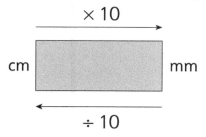

- Practise converting measures in centimetres to millimetres by multiplying by 10. Start with whole centimetres, then decimal fractions such as 3·5 cm, 4·2 cm. Now convert lengths in millimetres to centimetres by dividing by 10: 30 mm, 50 mm, 120 mm then 45 mm, 73 mm.

- Show children the metre stick. Ask: **How many centimetres in 1 m?** (*100*) Establish that their height is between 1 m and 2 m and that 10 cm is not even the length of a short ruler. Sketch a longer function machine, writing 'm' on the left and 'cm' on the right, with × 100 over the ruler and ÷ 100 underneath.

- Practise converting metres to centimetres by multiplying by 100: 2 m, 3·5 m, 4·25 m, 0·5 m. Then convert centimetres to metres by dividing by 100: 300 cm, 250 cm, 345 cm, 45 cm.

Teaching activity 2a (15 minutes)

Convert between g/kg, m/km and ml/litres

- Give children measuring containers and point out the measurements. Establish that 1000 g = 1 kg and 1000 ml = 1 litre. Ask: **What do we measure in grams and kilograms?** (*mass, how heavy something is*) **What do we measure in millilitres and litres?** (*capacity*) Show real-life examples. For example, say: **This bottles holds 500 ml; this packet holds 750 g of cereal.**

- Now discuss kilometres and metres. Ask: **What do we measure in metres and kilometres?** (*length and distance*) Say: **We still use miles to measure distance in our country, but in Europe you will see distances in kilometres.** Suggest that it would take at least 10 minutes to walk 1 km, giving an example such as from school to the local shops. Ask: **Do you think that 1 kilometre equals 10 m, 100 m or 1000 m?** (*1000*)

- Establish that the conversion factor for all these measures is 1000. Use a function machine diagram to convert between grams and kilograms, millilitres and litres, metres and kilometres, in both directions. Convert 3 km, 2·5 litres and 0.4kg; then convert 4000 g, 4500 ml, 4250 m, 300 g, 350 ml.

Teaching activity 2b (10 minutes)

To convert between g/kg, m/km and ml/litres

- Show children Resource 18: Conversion tables and use it to convert between grams and kilograms, millilitres and litres, and metres and kilometres. Choose from headings in the table below and write the **bold** numbers into the table on Resource 18: Conversion tables. Help the children to fill in the missing values, using doubling or halving and partitioning.

  ◆ 2 kg = double 1 kg = 2000g; 4000 g is double 2000 g

  ◆ 0·5 kg is half of 1 kg, so half of 1000 g = 500 g

  ◆ 2500 g = 2000 g + 500 g = 2 kg + 0·5 kg = 2·5 kg

kg/litre/km	gram/ml/metre
1	**1000**
2	2000
0·5	500
4	**4000**
2·5	**2500**
3·25	3250
0·4	**400**
0·45	**450**

Unit 35: Understand and use approximate equivalences between metric units and common imperial units such as inches, pounds and pints

Content reference domain: 5M6

Prerequisites for learning

Know common metric measures for length, mass and capacity

Learning outcomes

Convert between imperial and metric units of measure

Key vocabulary

Inch, foot, yard, ounce, pound, pint, gallon, mile

Resources

Resource 18: Conversion tables; Resource 19: Imperial measures; Resource 20: Approximate imperial and metric measure equivalents; milk cartons

Background knowledge

Imperial units are units of measurement that have been used since 1825 in many parts of the world. However, in the late 20th century, many countries started using the metric system, although some imperial units are still used in countries such as the UK and the USA. Most children still give their height and weight in imperial units and we still use miles for distance. Milk containers are still marked in pints with their metric equivalent shown either as ml or litres. Look at Resource 19: Imperial measures for more information.

Teaching activity 1a (20 minutes)

To convert between imperial and metric units of measure

- Explain that in the UK we use two different measuring systems, metric and imperial. Give a little of the history, if appropriate. Write the two headings on a whiteboard and ask children to supply as many units of measure as possible and to suggest whether they are metric or imperial. Include all the metric units covered in Unit 34 and the imperial units listed in the key vocabulary above. Explain that, because both systems are in use, it is helpful to know rough equivalents between the two systems.

- If possible, find a ruler or tape measure with inches on one side and centimetres on the other or use the image on Resource 19: Imperial measures. Ask: **How long is this ruler, in centimetres?** (*15 or 30*) **How many inches is that?** (*6 or 12*) Say: **This ruler shows that an inch is approximately 2·54 cm.** Use 1 inch = 2·5 cm to find approximate equivalences. Ask: **What is 2 inches in centimetres?** (*5 cm*) **10 inches?** (*25 cm*) **12 inches?** (*30 cm*) Say: **12 inches is also 1 foot, so 1 foot is approximately 30 cm. 3 feet make 1 yard, so how many centimetres are there in 1 yard?** (*90 cm, which is almost, but not quite, 1 metre.*)

- Repeat with scales, or image of scales, showing grams and kilograms, pounds and ounces, to find some equivalents. Discuss the markers on a pound and ounces scale, for example, asking: **What is each interval worth?** Explain: **There are 16 ounces in each pound, so each marker is worth 1 ounce.**

- Write several equivalents, from pounds to kilograms and grams, from kilograms and half-kilograms to pounds and ounces, to establish that 1 pound = 450 g and 1 kg = 2·2 pounds or 2 pounds 3 ounces. Ask: **What is 2 kg in pounds? What is 10 kg? How many grams make 2 pounds? How many kilograms in 10 pounds?** Ask the children to solve this problem: **A baby weighs 3 kg when he is born. What is that in pounds and ounces?** (*6 pounds 9 ounces or 6·6 pounds*)

- Repeat this type of activity with capacity jugs to show that 1 pint = 0·56 litres or 560 ml, 1 litre = $1\frac{3}{4}$ pints. Ask: **How many litres in 2 pints? 4 pints?** Check with the milk containers, which are accurate to the nearest millilitre. Ask: **How accurate are our conversions?**

- Explain that some other metric–imperial equivalents are shown on Resource 19: Imperial measures. Use them to work out other approximate equivalents.

  Note: Do not use 5 miles = 8 km or 1 fluid ounce = 30 ml as these are on Resource 20: Approximate imperial and metric measure equivalents.

Teaching activity 1b (20 minutes)

To convert between imperial and metric units of measure

- Remind children that two different measuring systems are used in the UK; these are metric and imperial. Give a brief history, if appropriate. Write the two headings on a whiteboard. Ask children to supply as many units of measure as they can. Discuss whether each is metric or imperial and list them in appropriate columns. Include all the metric units covered in Unit 34 and the imperial units listed above, in the key vocabulary. Explain the importance of knowing the **approximate** equivalents.

- Choose one conversion, for example, 1 inch = 2·54 cm. Explain that we only really need an approximate equivalent, so we can use 2·5 cm. Use Resource 18: Conversion tables and refer to the information on Resource 20: Approximate imperial and metric measure equivalents to list some conversions. Write 'inches' in the top left row and 'cm' in the top right row; in the second row put the conversion: 1 under inches and 2.5 under cm. Work out the centimetre equivalents for 2, 5, 10, 12, 24 and 36 inches, using known relationships. For example, say: **2 inches is double 2·5 cm, which is 5 cm. 10 × 2·5 = 25 cm, so 5 inches is half of 25 = 12·5 cm. 12 inches = 10 + 2 inches so 30 cm.**

- Explain that 12 inches make 1 foot, so we can see that 1 foot is about 30 cm. Say: **You can work out other equivalents by partitioning into the values shown in the table, for example, 4 feet and 8 inches in cm is 4 × 30 + 25 − 5 = 140 cm = 1 m 40 cm.**

- Work out other heights given in feet and inches. Use the information in the table to convert from metric to imperial: 1 m is 90 cm + 10 cm = 3 × 12 + 4 inches = 40 inches.

- Complete a similar table for mass. You could suggest that children use 1 ounce = 30 g for easier numbers. Explore 16 ounces = 1 pound or 14 pounds = 1 stone. Ask: **What is 3 pounds in grams?** *(1350 g or 1·35 kg)* **What is 7 stone in kg?** *(about 44 kg)* Use the information in the table to convert from metric to imperial, using partitioning into multiples of 30 g; ask: **What is 200 g in ounces?** *(210g = 7 × 30 g so 200 g is approximately just under 7 ounces.)*

- Complete a similar table for capacity. Ask questions related to the conversion they have used. Say: **1 pint = 0·54 litres: 1 gallon is 8 pints.** Ask: **How many litres in 1 gallon? Why is this not the same as on the RS?** *(The conversions are approximate.)* Say: **1 gallon = 4·5 litres, 1 gallon is 8 pints, how many millilitres make 1 pint?**

- Look at some more of the conversions on Resource 20: Approximate imperial and metric equivalents that you have not used and ask questions, without using a table, such as: **1 gallon is 4·5 litres, so how many litres in 2 gallons? In 4 gallons? In 10 gallons? A recipe needs 100 g of sugar. Approximately how many ounces is this? The door is 2 m high, how tall is it in feet and inches?**

Unit 36: Measure and calculate the perimeter of composite rectilinear shapes in centimetres and metres

Content domain reference: 5M7a

Prerequisites for learning

Count squares to measure perimeter

Use a ruler to measure accurately in centimetres

Learning outcomes

Calculate perimeter by measuring then adding dimensions

Key vocabulary

Perimeter, dimension, accurate

Resources

Resource 21: Perimeter and area; 1 cm squared paper; rulers

Background knowledge

Perimeter is the distance around a 2-D shape. It is the total length of the lines that make the outline of the shape. This lesson moves on from counting squares, to beginning to use a formula to find the perimeter of shapes such as squares and rectangles. It also looks at how to find the perimeters of shapes made up of two or more squares or rectangles (**composite rectilinear** shapes).

Teaching activity 1a (20 minutes)

To calculate perimeter by measuring then adding dimensions

- Ask children: What is the perimeter of a shape? (*a measure of the length of the outside edges of a shape*) Draw a square on a whiteboard, labelling each side 5 cm. Explain that this is not drawn to the correct size, so the sides are not actually 5 cm long. Ask: **How can we work out the perimeter of this square?** (*Add up the lengths of all the sides.*) Record this as $P = 5$ cm $+ 5$ cm $+ 5$ cm $+ 5$ cm $= 20$ cm. Ask: **How could we write this as a multiplication?** ($P = 4 \times 5$ *cm* $= 20$ *cm*)

- Draw a rectangle, labelling the sides: 5 cm, 3 cm, 5 cm, 3 cm. Ask: **How can we work out the perimeter of this rectangle?** (*Add up the lengths of all the sides.*) Record this as $P = 5$ cm $+ 3$ cm $+ 5$ cm $+ 3$ cm $= 16$ cm. Ask: **How could we write this using multiplication?** ($P = 2 \times 5 + 2 \times 3 = 10 + 6 = 16$ *or* $P = 2 \times (5 + 3) = 2 \times 8 = 16$)

- Say: **You are going to explore the perimeters of squares and rectangles by drawing them on squared paper. You will write the perimeter as an addition and then by using multiplication.** Ask the children to draw squares of side 2 cm, 3 cm and 4 cm, and rectangles measuring 2 cm by 3 cm, 3 cm by 4 cm, 4 cm by 5 cm. Ask them to record the perimeters, as explained above.

- Demonstrate drawing a composite shape, as on Resource 21: Perimeter and area, labelling each dimension in single-digit **metre** measures, for example, 8 m, 6 m, 6 m, 5 m, 3 m, 3 m and 3 m. Say: **This is the shape of a farmer's paddock. He wants to put up a perimeter fence all around his paddock. How many metres of fencing will he need?** Record the perimeter as an addition, for example: $8\text{ m} + 6\text{ m} + 6\text{ m} + 5\text{ m} + 3\text{ m} + 3\text{ m} + 3\text{ m} = 34\text{ m}$. Ask: **Have we included every line?**

- Children explore drawing shapes made from two or more rectangles and/or squares and working out the perimeter, in centimetres. Watch out for children just working out the perimeters of the two shapes and adding them together. Sides that join one shape to another are not included in the perimeter as they are inside the shape.

- For extra practice, and to check understanding, label a copy of Resource 21: Perimeter and area with lengths in centimetres or metres, or a mixture. Add the note: 'Not drawn to scale', before reproducing it for each child. Children find the new perimeters by adding the given dimensions.

Note: This is how SATs questions will be presented.

Teaching activity 1b (20 minutes)

Calculate perimeter by measuring then adding dimensions

- Revise the definition of perimeter with the children. Then draw a square of side length 5 cm and a rectangle measuring 5 cm by 3 cm and work out the perimeters, recording them in two ways. For the square: $P = 5\text{ cm} + 5\text{ cm} + 5\text{ cm} + 5\text{ cm} = 20\text{ cm}$ and $P = 4 \times 5\text{ cm} = 20\text{ cm}$. For the rectangle: $P = 5\text{ cm} + 3\text{ cm} + 5\text{ cm} + 3\text{ cm} = 16\text{ cm}$ and $P = 2 \times 5 + 2 \times 3 = 10 + 6 = 16\text{ cm}$ or $P = 2 \times (5 + 3) = 2 \times 8 = 16\text{ cm}$.

- Give each child Resource 21: Perimeter and area. Explain that they should measure each side **to the nearest centimetre** and label each one, in cm. Ask them to work out the perimeters, recording them underneath each shape, first as additions and then using multiplication. Children measure and record perimeters as above. for the three squares and three rectangles.

- Direct children to look at the first compound shape on Resource 21: Perimeter and area, made from one rectangle and one square. Ask them to measure the different lengths, record the dimensions on the figure and use addition to work out the perimeter. Repeat with the next shape. Continue, until children are confident.

- In some shapes it is clear that two opposite sides are equal, but it is best at this stage simply to add up the dimensions of all the sides. In Year 6 they will be presented with shapes in which not all sides are labelled. They will be expected to work out missing dimensions before working out the perimeter.

- For extra practice, and to check understanding, provide further composite shapes, or label a copy of Resource 21: Perimeter and area with lengths in centimetres or metres, or a mixture. Add the note: 'Not drawn to scale', before reproducing it for each child. Children find the new perimeters by adding the given dimensions.

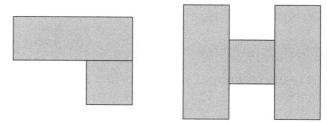

Note: This is how SATs questions will be presented.

Unit 37: Calculate and compare the area of rectangles (including squares), including using standard units, square centimetres (cm²) and square metres (m²), and estimate the area of irregular shapes

Content domain reference: 5M7b

Prerequisites for learning

Know how to work out and record square numbers, using index notation

Work out area by counting squares

Learning outcomes

Use multiplication to work out the areas of squares and rectangles

Estimate the area of irregular shapes by dividing them into rectangles

Key vocabulary

Area, cm², m², square metres, square centimetre, estimate, dimension, width length

Resources

Resource 21: Perimeter and area; 1 cm squared paper; rulers; medium–large 2-D shapes: circles, triangles, pentagons, hexagons, trapeziums, parallelograms; four metre rules

Background knowledge

Area is the amount of space **inside** a 2-D shape. Area is measured in square units, for example, square centimetres (cm²), square metres (m²), square miles. The most efficient way to work out area is to multiply the values of the two dimensions, width and length.

When estimating the area of an irregular shape, partial squares have to be considered. This lesson explores several strategies.

Teaching activity 1a (20 minutes)

Use multiplication to work out the areas of squares and rectangles

- Define area as the space contained inside a 2-D shape. Discuss the units of area: square centimetres, square metres, square kilometres, square miles. Show children a square centimetre by shading in one square on a sheet of centimetre square paper. Show them a square metre by laying out the four metre sticks in a square.

- Draw a square on a whiteboard, labelling each side 5 cm. Explain that is not drawn to scale, so is each side is not actually 5 cm. Ask: **How can we work out the area of this square?** Within the shape, draw lines to show five rows of five squares and count the rows in 5s, saying: **5, 10, 15, 20, 25: the area of the square is 25 square centimetres. We can write this using index notation**. Review how they wrote square numbers. Write: *A* (area) = 25 cm². Ask: **Is there a quicker way than counting in 5s?** (*There are 5 squares in every row and there are 5 rows, so 5 × 5 = 25.*) Say: **We can multiply the number of rows by the number of columns.**

- Draw a rectangle, labelling the sides, in order, 5 cm, 3 cm, 5 cm and 3 cm. Within the rectangle, draw lines to show three rows of five squares. Ask: **How can we work out the area of this rectangle?** (*Counting in 5s: 5, 10, 15, 15 cm² or multiplication: 5 × 3 = 15 cm².*) Record, using multiplication: *A* = 5 × 3 = 15 cm².

- Children now find the areas of squares of side 2 cm, 3 cm and 4 cm and rectangles measuring 2 cm by 3 cm, 3 cm by 4 cm, 4 cm by 5 cm by drawing them on squared paper. They record the areas, using the correct notation for square centimetres (cm²). Encourage them to multiply the dimensions rather than counting squares.

- Extend, if appropriate, by drawing shapes made of two or more rectangles and working out the total areas of these composite shapes by splitting them into rectangles or by exploring squares, to 12 by 12, and rectangles with consecutive dimensions, 6 by 7, 7 by 8…

- For further practice, label the squares and rectangles on Resource 21: Perimeter and area with whole-number lengths, then give children copies, explaining that these are not the actual sizes, so they should just use the numbers given.

Teaching activity 1b (20 minutes)

Use multiplication to work out the areas of squares and rectangles

- Define area as the flat space taken up a 2-D shape, showing what 1 square centimetre and 1 square metre look like.

- Sketch a 5 cm square. Show children how to use multiplication to work out the area: five rows of five squares is $5 \times 5 = 25$ squares. Tell the children that they can use index notation for square centimetres, as they did for square numbers: 25 cm^2.

- Repeat for a 3 cm by 5 cm rectangle, counting five rows of three squares (5×3) or three rows of five squares (3×5): $5 \times 3 = (3 \times 5 =) 15$ squares. The area is 15 square centimetres or 15 cm^2.

- Children use a ruler to measure the sides of the squares and rectangles on Resource 21: Perimeter and area (or use the dimensions measured in Unit 36 activity 1b) to work out the areas, writing the calculations and the answers in square centimetres (cm^2).

- Continue by working out the areas of the composite shapes by splitting them into two or three rectangles. Demonstrate the first one if necessary, showing that it does not really matter which way the shape is split, the total area will be the same.

Teaching activity 2a (20 minutes)

Estimate the area of irregular shapes by dividing them into rectangles

- Lay your hand, palm down and fingers closed, on squared paper. Draw around it. Show children how to split the outline into rectangles, each as large as possible and including rectangles with a width of 1 cm around the edges where possible.

Work out the area of each rectangle, add them to produce a total for the whole squares. For the partial squares, **either** count partial squares that are larger than a half square, ignoring the squares that are less than a half **or** combine each larger partial square with a smaller one to make whole squares. Add these to the previous total of whole squares, to find the grand total area.

- Children work out an estimate for the area of their own hands. If there is time, allow them to find the approximate area of their feet or circular objects.

Teaching activity 2b (20 minutes)

Estimate the area of irregular shapes by dividing them into rectangles

- Draw an irregular shape on squared paper, not too large and roughly circular. Show children how to draw as large a rectangle as possible inside the shape and as small a rectangle as possible around the outside of the shape. Work out the areas of the two rectangles, explaining that the area of the shape must be between these two areas – smaller than the outside one, but larger than the inside one. Ask: **What is a good estimate for the area of this shape?**

- Children draw around 2-D shapes, such as circles, triangles and pentagons, on centimetre-squared paper, drawing outside and inside the shape to work out its approximate area. Show them how to align any straight edges of a 2-D shape to a grid line, to ensure a more accurate estimate. They could also use small classroom objects, such as a reel of sticky tape, the circular base or rim of a plant pot, or they could draw their own shapes.

- Alternatively, they could split 2-D shapes and objects into rectangles and count the squares and partial squares.

- As an extra activity, children explore triangles to discover that a triangle's area is half that of the rectangle that it surrounds it.

Unit 38: Estimate volume and capacity

Content domain reference: 5M8

Prerequisites for learning

Express cube numbers in index notation
Round to the nearest 100 ml or litre

Learning outcomes

Work out volumes of cuboids
Estimate capacity

Key vocabulary

Capacity, volume, cube, cuboid, estimate

Resources

Resource 22: Cuboids; centimetre interlocking cubes; isometric or triangular grid paper; small cuboid containers; litre measuring jug, variety of different sized containers for liquids

Background knowledge

The **capacity** of a container is the maximum amount it can hold. **Volume** is the amount of space taken up by a 3-D object. A jug may have a capacity of 500 ml, but the volume of the milk in that jug can vary. To fill it to capacity requires a volume of 500 ml of liquid, such as milk.

Volume and capacity are measured in cubic centimetres (cm^3), cubic metres (m^3) and litres.

Teaching activity 1a (15 minutes)

Work out volumes of cuboids

- Ask children to use interlocking cubes to show 2^3, as in Unit 18, activity 2a: they will need two layers of four cubes. Say: **This cube has a volume of 8 cubic centimetres, which we write like this: 8 cm^3.** Ask children to make a cube to show 3^3. They will need three layers of nine cubes. Ask: **What is the volume of this cube?** (*27 cubic centimetres*) **How do you write this in index notation?** (*27 cm^3*) Remind children that the cube numbers, and therefore the volumes of cubes, are made by multiplying the length by the width by the height. For the cube of side 3 cm, this means $3 \times 3 \times 3$ as all the dimensions are the same. (This is true for all cubes.) Ask: **What is the volume of a cube of length 4 cm?** ($4 \times 4 \times 4 = 64\ cm^3$)

- Build some cuboids together. Count the cubes to work out the volumes and record them, in cm^3, on a slip of paper. Turn the cuboid so the base becomes the height and repeat. Ask: **Is the volume the same? Can you see a way to use multiplication to work out the volume?** Make the link with area: **The bottom layer is made up of two rows of three, so that is six cubes. There are four layers, so altogether, $6 \times 4 = 24$ cubes. The volume is 24 cm^3.**

- Explain that this is the formula or rule for working out the volume of a cuboid. Say: **You multiply the length by the width then multiply the answer by the height.** Check, using this formula, that their counting was correct when they built some cuboids.

- Build some more cuboids and work out the volume each time, just using multiplication. Then count to check that the answer is correct.

Teaching activity 1b (15 minutes)

Work out volumes of cuboids

- Demonstrate how to use isometric paper to draw cubes and cuboids (see the diagram below). Join the dots to show the cubes. Work out the volume of each cube or cuboid by counting the cubes, including the 'hidden' ones. Children draw two more cubes and some cuboids and work out their volumes, recording them in cm^3. Alternatively, let them work out the volumes of the cuboids on Resource 22: Cuboids.

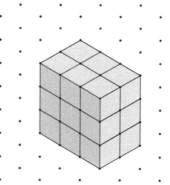

- Demonstrate how to work out the number of centimetre cubes in one layer, for example, for a 2 by 3 by 3 cuboid, the base is 2 by 3 = 6 centimetre cubes, then multiply that by the number of layers (3), to give width × length × height, in this case 2 × 3 × 3 = 18. The volume is 18 cm³. Explain that this is the formula or rule for working out the volume of a cuboid. Check children's understanding by drawing a cuboid with dimensions marked, but no cubes showing. Ask: **What is the volume of this cuboid?** (*10 × 5 × 3 = 150 cm³*)

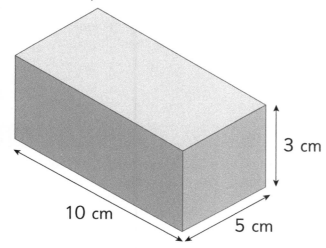

3 cm

10 cm

5 cm

Teaching activity 2a (15 minutes)

Estimate capacity

- Collect six to eight differently shaped watertight containers. Put 500 ml of water into a 1 litre jug and tell the children: **This is 500 ml.** Ask the children to separate the containers into those they think will have a capacity of less than 500 ml and those that have a capacity of more than 500 ml. Pour water into some of the containers to check. Each time, ask: **Is this container full to capacity?** They order the containers from smallest capacity to largest and make a note for later.

- Empty all the containers and refill the jug to 500 ml. Starting with the container the children thought was the smallest, fill it with as much of the water from the jug as possible and put the jug with the remaining water next to the container. Without children looking at the measuring scale on the jug ask: **Approximately, what is the capacity of this container?** (*It is less than 500 ml.*) **But is there more left in the jug or more in the container?** Children write their estimate, then work out the actual amount by reading what is left in the jug and subtracting it from 500 ml. Work through the containers less than 500 ml from smallest guessed capacity, to largest, ensuring there is 500 ml of water in the jug each time. Re-order in capacity size if necessary.

- Repeat with the larger containers. Pour all 500 ml of the water from the jug into each container in turn, asking children: **Now can you guess the total capacity of the container?** Fill the container up and then pour all the water back into the measuring jug to measure its actual capacity. Ask: **How close was your estimate?**

- Ask: **Did you order the containers correctly? Were there any surprises? Why is it difficult to estimate which containers hold more than others when they are all different shapes?** (*Tall thin containers look as if they will hold more than short, wide ones.*)

Teaching activity 2b (15 minutes)

To estimate capacity

- Show the children three to six cuboid containers and the centimetre cubes. Ask: **How many of these cubes will fit into each container?**

- Explain that these are centimetre cubes, so the number of cubes that will fit in without spaces between them is the capacity of the container, in cubic centimetres.

- Write the children's estimate of its capacity on each container. Then ask the children to fill each one by tipping the cubes into it and shaking them down. They then tip out the cubes and count them to work out the approximate capacity of the container. Ask: **How close was your estimate?**

- If time, they could place the cubes into the containers carefully, so that there are no gaps, to work out the actual capacity or use the cubes to show the dimensions of the container and use multiplication to find the volume.

Unit 39: Solve problems involving converting between units of time

Content domain reference: 5M4

Prerequisites for learning

Convert between weeks, days, hours, minutes and seconds

Read, write and convert time between analogue and 24-hour digital clocks

Learning outcomes

To convert between units of time

To work out durations of time

Key vocabulary

Analogue, digital, 24-hour clock

Resources

Resource 23: Analogue and digital clock faces; clocks with movable hands; teaching clock; A4 plain paper

Background knowledge

Our system of time is based on an ancient Babylonian system which was based on 60 and is linked to the movement of the Earth. The Earth rotates on its axis approximately every 24 hours, causing day and night. It also orbits the Sun every 365.25 days which gives us seasons. A calendar year has 365 days for three years and 366 days every fourth or leap year, to catch up. The numbers of leap years are divisible by 4. Children should be given experience of both types of clocks and watches as both are used extensively in different contexts.

Teaching activity 1a (20–30 minutes)

Convert between units of time

- Children make an A4 revision poster to show:

  1 year = 12 months = 365 days

  1 week = 7 days; 1 day = 24 hours

  1 hour = 60 minutes; 1 minute = 60 seconds

  months with 31 days: January, March, May, July, August, October, December

  months with 30 days: April, June, September, November

  February: 28/29 days

- Children use the poster as a support to answer questions, such as: **How many days in 6 weeks?** (*42 days*) **How many months in 3 years** (*36 months*) **How many seconds in $4\frac{1}{4}$ minutes?** (*255 minutes*) **How many hours in 5 days?** (*120 hours*) **How many days from 1 September to New Year's Eve?** (*122 days*) Ask children to suggest more questions to ask each other.

- Extend the questions: **How many minutes are there in 1 day?** (*1440 minutes*) **How can we use 365 days = 1 year to work out how many weeks there are in a year?** (*Use division to prove that there are 52 weeks in a year: 365 ÷ 7 = 52 weeks 1 day.*) **How many hours and minutes is 135 minutes?** (*135 ÷ 60 = 2 remainder 15, so 2 hours 15 minutes.*) It is important that they interpret the remainder correctly. A decimal remainder is not appropriate here.

- If appropriate, discuss the history and science of our system of time.

Teaching activity 1b (20 minutes)

Convert between units of time

- Recap time conversions relating to seconds, minutes, hours, days, weeks, months and years. Ask: **How many minutes are there in 1 hour?** (*60*) **How many days are there in a week** (*7*) **How many days are there in a leap year?** (*366*)

- Pose the question: **Have you or your teacher been alive for 1 million hours?** Ask children to write 'yes' or 'no', as their immediate response, on a slip of paper and hide it.

- Ask: **How can we work out how many hours we have been alive? Does it need to be absolutely accurate? What time facts do we need to know to solve this problem?** Write some of them down: 365 days in a year, 24 hours in a day…

- Ask: **How old are you? How many days is that?** (*9 × 365 or 10 × 365*) For absolute accuracy they could add the extra days from their last birthday and the extra days for the number of leap years they have been alive to the total number of days. Ask: **What do we need to do next?** (*Multiply by 24 to give the number of hours.*)

- The answer is approximately 78 900 (for a 9-year-old child), 87 600 (for a 10-year-old child) and 96 400 (for 11 year olds). Ask: **What about me? Have I been alive 1 million hours? Instead of multiplication we can use division, but we will need a calculator: 1 000 000 hours divided by 24 is about 41 666 days, then divide by 365 to get 114 years, which is longer than even a teacher can have been alive!**

- Pose other similar questions: **How old is a child that has been alive 1 million seconds?** (*about 11 days*) **A million minutes?** (*about 21 months*)

Teaching activity 2a (15 minutes)

Work out durations of time

- Use Resource 23: Analogue and digital clock faces. Revise converting between analogue and digital times, using quarter to three, ten past five, twenty-five to two. Show the correct positions of the hands on the analogue clock face, writing the time in words underneath. Show the two possible matching digital times on the digital clocks, labelling them as a.m. and p.m. Discuss the differences between the analogue and digital faces.

  - Clock 1: quarter to 3 = 2:45,14:45, 2.45 a.m., 2:45 p.m.
  - Clock 2: ten past five = 5:10, 17:10, 5:10 a.m., 5:10 p.m.
  - Clock 3: twenty-five to two = 1:35; 13:35, 1:35 a.m., 1:35 p.m.

- Use the time on the clocks to ask questions about duration or time intervals: **How many hours and minutes have passed between the times on these two clocks?**

  - 1 to 2: 2 hours 25 minutes
  - 3 to 2: 3 hours 35 minutes
  - 3 to 1: 1 hour 10 minutes

- Point to one of the clock faces or show a time on a teaching analogue clock. Say:
  - **This is the time shown on Aled's watch. His watch is 12 minutes slow, what time is it?**
  - **Pasha's watch is 6 minutes fast. What time is it?**
  - **The bus is due in 11 minutes. What time should it arrive?**
  - **The station clock shows 2:45. The next train arrives at 4:05. How long is it before the next train?**

- Repeat with similar questions, using some 24-hour digital times written on a whiteboard.

Teaching activity 2b (15 minutes)

To work out durations of time

- Use the teaching clock to revise telling the time to the nearest minute and then converting this to the two possible digital times. Point out that the difference in the number of hours is 12, but the minutes stay the same. Explain that times before 10:00 a.m. should be written with a preceding zero, 9:24 = 09:24. 9.24 in the evening would be 21:24 (9 + 12 = 21).

- Show different times on two clock faces, one to three hours apart, by multiples of 5 minutes but **not** whole or half hours, unless this is a necessary starting point. Say: **These are a.m. times. What are the two digital times?** Ask appropriate questions related to the duration between the times on the two clocks. Say: **This is the time Luke arrived at his friend's house; this is the time he left. How long was Luke at his friend's house?** Then: **These are the start and finish times of a film. How long was the film?** Children can use clocks with movable hands as support.

- Use one clock at a time, switching between analogue and digital, to ask duration questions.
  - **It is 20:50. Alex will go to bed in 40 minutes. What time will he go to bed?**
  - **A film is 2 hours and 40 minutes long. It ended at 10:15. What time did it start?**
  - **Mae takes 35 minutes to walk to school. What time will she need to set off to get to school by ten to nine?**

- Extend the activity by repeating with longer durations; use times that are not multiples of 5 minutes. Constantly revise the conversion between analogue and 24-hour digital times.

Unit 40: Use all four operations to solve problems involving measure using decimal notation, including scaling

Content domain reference: 5M9

Prerequisites for learning

Convert between metric units of mass, length and capacity

Find fractions of amounts

Find the value of one, given the value of many

Learning outcomes

Use scaling by fractions to solve problems involving measure

Use decimal notation to solve problems involving measure

Solve simple problems involving proportion (scaling)

Key vocabulary

Mass, length, capacity, proportion

Resources

Resource 18: Conversion tables; Resource 10: Decimal place-value grid 1 (optional)

Background knowledge

Many of the concepts in this unit have been covered in previous measurement units. The skills learned must be applied to solve problems in this unit. If children are still struggling with the basic concepts, revise the relevant units before attempting these activities.

Scaling is covered in these activities in the contexts of fractions of measures and in solving proportion problems.

This is the only Year 5 unit that covers using all four operations to solve problems involving money and decimals.

Teaching activity 1a (20 minutes)

Use scaling by fractions to solve problems involving measure

- Use Resource 18: Conversion tables, to work out grams and centimetres as fractions of kilograms and metres respectively. Write 'kg' and 'grams' as headings in the first table. In the next row write the equivalents: 1 (kg) = 1000 (g). Write $\frac{1}{2}$ in the kg column and ask children: **How many grams are there in half a kilogram?** (*500 g*) Repeat for a quarter, halving the half. Ask: **How many grams are there in three-quarters of a kilogram?** $\frac{3}{4} = \frac{1}{2} + \frac{1}{4}$ or $3 \times \frac{1}{4}$, so $\frac{3}{4}$ kg = 750 g. Continue with $\frac{1}{10}$ ($\div$ 10) = 100 g, $\frac{1}{5}$ (double 1 tenth) = 200 g. Then use these to find multiples of tenths and of fifths to fill the table.

- Children use the second table to work out the same fractions of metres in centimetres. Start them off with 1 m = 100 cm and $\frac{1}{2}$ m = 50 cm. Then $\frac{1}{4}$ m = 25 cm, $\frac{3}{4}$ m = 75 cm, $\frac{1}{10}$ m = 10 cm, $\frac{1}{5}$ m = 20 cm, $\frac{3}{10}$ m = 30 cm, $\frac{3}{5}$ m = 60 cm.

- Draw attention to the link with fraction, decimal and percentage equivalents.

- Ask and work through related questions: **How many millilitres are there in two-fifths of a litre?** (*400 ml*) **A bottle contains a quarter of a litre of water. How many millilitres is that?** (*250 ml*) **A recipe needs half a kilogram of flour. How many grams of flour is that?** (*500 g*) **What is 75 cm as a fraction of 1 metre?** ($\frac{75}{100} = \frac{3}{4}$) **Al lives three-quarters of a kilometre away from school. How many metres is that?** (*750 m*) **Write $2\frac{1}{2}$ kg in grams.** (*2500 g*) **Write $1\frac{3}{4}$ litres in millilitres.** (*1750* ml) **How many 200 ml servings can I get from a 2 litre bottle?** (*200 ml is a fifth of 1 litre, so 5 servings from a litre, 10 serving from 2 litres.*) If it is helpful, use more tables on Resource 18: Conversion tables to work out the answers.

Teaching activity 1b (20 minutes)

Use scaling by fractions to solve problems involving measure

- Draw number lines to show quarters. Write 0 kg = 0 g on the left and 1 kg = 1000 g on the right. Ask: **Which two values do I need to write halfway?** ($\frac{1}{2}$ kg = 500 g) Repeat for $\frac{1}{4}$ kg = 250 g and $\frac{3}{4}$ kg = 750 g.

- Draw a number line in tenths, or use Resource 18: Conversion tables. Repeat for tenths: $\frac{1}{10}$ kg = 100 g, $\frac{2}{10}$ = 200 g, $\frac{9}{10}$ kg = 900 g . Write the fifths underneath: $\frac{1}{5}$ kg = 200 g and so on.

- Repeat with a number line in tenths for 1 metre = 100 cm: $\frac{1}{2}$ m = 50 cm, $\frac{1}{4}$ m = 25 cm, $\frac{3}{4}$ m = 75 cm; $\frac{1}{10}$ m = 10 cm, $\frac{2}{10}$ m = 20 cm… $\frac{9}{10}$ m = 90 cm; $\frac{1}{5}$ m = 20 cm; $\frac{2}{5}$ = 40 cm, $\frac{3}{5}$ m = 60 cm, $\frac{4}{5}$ = 80 cm.

- Work out $\frac{1}{1000}$ and $\frac{1}{100}$ of 1 kg (1 g, 10 g) and $\frac{1}{100}$ of 1 metre (1 cm).

- Ask questions relating to fractions of measures. **What is $2\frac{1}{5}$ metres in centimetres?** (220 cm) **How many 50 cm lengths of string can be cut from a 4 m roll?** (8)

Teaching activity 2a (20 minutes)

Use decimal notation to solve problems involving measure

- Write 1.234 on a whiteboard and ask children to give the value of each digit: 1 whole, 2 tenths, 3 hundredths, 4 thousandths. Write kg after the number: 1.234 kg. Ask: **What is the value of the 2 now?** ($\frac{2}{10}$ of 1 kg: 1 kg = 1000 g, so $\frac{2}{10}$ is 2 × 100 g = 200 g.) **What is the value of the 3 now?** ($\frac{3}{100}$ of 1 kg: 3 × 10 g = 30 g) **What is the value of the 4 now?** ($\frac{4}{1000}$ of 1 kg: 4 × 1 g = 4 g) **So how many grams altogether is 1.234 kg?** (1000 + 200 + 30 + 4 = 1234 g) **Can you see the connection? What is 1.23 kg in grams?** (1000 + 200 + 30 = 1230 g) **What is 1.2 kg in grams?** (1000 + 200 = 1200 g) Explain that zeros are used as place holders to make a four-digit number.

- Repeat for other amounts, asking: **What is 2.468 litres in millilitres? What is 2.46 litres? What is 2.4 litres? What is 4.276 km in metres? What is 4.2 km? What is 4.2 km?**

- Write 0.583 km on a whiteboard. Ask: **What is 0.583 km in metres?** (500 m + 80 m + 3 m = 583 m) Then let children try expressing 0.352 kg in grams, 0.474 litres in millilitres, then 0.87 kg in grams (800 + 70 = 87**0**) and 0.3 km in metres (300 m).

- Repeat for other amounts: **What is 5.67 metres in centimetres?** (500 cm + 60 cm + 7 cm = 567 cm; 5.6 m = 500 cm + 60 cm = 560 cm) Ask children to try 6.38 m and 7.09 m.

- Repeat for 0.45 metres (40 + 5 = 45 cm) and 0.9 metres (90 cm). Then ask: **What is 1 kg 456 grams in kilograms?** (1.456 kg) **What is 3 km 450 m in kilometres?** (3.45 km)

- Finally, set some measures-related problems: **Which is more, 1.76 kg or 1749 grams?** (1.76 kg) **Lucy walks 3.64 km to school, then walks home again. How far does she walk each day?** (7.28 km) **The mass of 1 coin is 6.5 grams. What is the mass of 100 of these coins in kilograms?** (650 = 0.65 kg) **How many 250 ml servings can be made from a 2.5 litre bottle?** (10)

Teaching activity 2b (20 minutes)

Use decimal notation to solve problems involving measure

- Write **1 kg 275 g** on a whiteboard. Ask: **How many grams is this?** (1000 + 275 = 1275 g) **How can we write this in kilograms?** (Divide the grams by 1000: **1.275** kg.) Use a place-value grid, such as Resource 10: Decimal place-value grid 1, if necessary.

- Ask children to write some other amounts in grams and in kilograms and repeat, converting to grams and to kilograms as a decimal.

- Repeat for **1 kg 450 g** and **2 kg 600 g**, noting how the kilogram decimal is written: 1450 g and 1.45 kg, 2600 g and 2.6 kg. Children try 2 km 438 m, 2 litres 750 ml; 5 km 300 m…

- Now try the reverse, asking: **What is 3.76 km in kilometres and and metres? What is 4867g in kilograms and grams? What is 5980 ml in litres? What is 4.5 litres in millilitres?**

- Set a word problem: **Al runs 9.4 km, Chris runs 7546 m. How much further did Al run than Chris? What do we need to do to solve this problem?** (Subtract 7546 m from 9.4 km.) **What do we need to do first?** (Change kilometres to metres or metres to kilometres, so that the units are the same for both runners.) Work it out both ways: 9.4 – 7.456 = 1.944 km (recap using zeros as place holders to make the decimal parts the same length and exchanging, Unit 31) and 9400 – 7456 = 1944 m = 1.944 km.

Teaching activity 3a (20 minutes)

Solve simple problems involving proportion (scaling)

- Recap simple rates (Unit 21), asking: **If 10 pens cost £2.40, what does 1 pen cost? What do 5 pens cost?** (*240 divided by 10 = 24p so 1 pen costs 24p.*)

- Write this list of ingredients on a whiteboard: 120 g sugar, 240 g butter, 300 g flour, 60 ml milk. Explain that this recipe makes 12 biscuits. Ask: **How much butter is needed to make 24 biscuits?** (*24 is double 12, so we need double 240 g, so 480 g.*) **How much flour is needed for 6 biscuits?** (*6 is half of 12, so we need half the amount of flour, half of 300 g is 150 g.*) **A school cook increases all the ingredients to make lots of biscuits. She uses 600 ml of milk. How many biscuits did she make?** (*600 ml = 10 × 60 ml, so we need to multiply the number of biscuits by 10 as well: 10 × 12 = 120, therefore she made 120 biscuits.*) **How many kilograms of sugar/butter/flour did she use?** (*12 × 120 g = 1440 g = 1.44 kg; 12 × 240 = 2880 g = 2.88 kg; 12 × 300 g = 3600 g = 3.6 kg*)

- Work out the amount of each ingredient needed for 1 biscuit by dividing by 12 (10 g sugar, 20 g butter, 25 g flour, 5 ml milk), then use this to work out the amount of butter for 5 biscuits (5 × 20 g = 100 g), sugar for 15 biscuits (15 × 10 = 150 g), flour for 8 biscuits (8 × 25 g = 200 g) and milk for 7 biscuits (7 × 5 = 35 ml).

Teaching activity 3b (20 minutes)

Solve simple problems involving proportion (scaling)

- Begin as activity 3a, recapping on simple rates (Unit 21). Write the ingredients for the recipe from the previous activity on the board: 120 g sugar, 240 g butter, 300 g flour, 60 ml milk.

- Explain that people often want to increase the recipe to make more biscuits, or decrease it to make fewer. If someone wanted to make 24 biscuits, they would not make 12 biscuits, then another 12 biscuits. Because 24 is double 12, they would need double the amount of each ingredient. Emphasise that each ingredient needs to be increased in proportion, otherwise the recipe would not work – the ingredients would not be in the correct ratio to each other.

- Use Resource 18: Conversion tables, with a different table for each ingredient. Work out together the amount of sugar needed for 24, 6, 3 and 18 biscuits, using doubling and halving and then dividing by 12 for 1 biscuit, and completing the table for 18 biscuits from the information they already have in the table (12 + 6 so 1 and a half lots). They do the same for 15 biscuits: (12 + 3) and 10 biscuits: 10 × 1 biscuits. Explain that they can now work out the amount of sugar for any number of biscuits as they know that each biscuit needs 10 g of sugar.

Biscuits	Sugar(g)
12	120
24	240
6	60
3	30
1	10
18	180
15	150
10	100

- Children complete the tables for butter, flour and milk with your support.

Biscuits	Butter(g)
12	240
24	480
6	120
3	60
1	20
18	360
15	300
10	200

Biscuits	Flour(g)
12	300
24	600
6	150
3	75
1	25
18	450
15	375
10	250

Biscuits	Milk(ml)
12	60
24	120
6	30
3	15
1	5
18	90
15	75
10	50

• Ask children to use the information in the tables to answer questions: **How much butter will be needed to make 8 biscuits? Sugar for 20 biscuits? Milk for 9 biscuits? Flour for 16 biscuits? If Rani uses 360 g of sugar, how much butter, flour and milk does she use? Cook used 600 ml of milk to make biscuits for lunch. How many biscuits did she make?** ($10 \times 12 = 120$, see activity 3a) **Max used 300 g of sugar, how much butter did he use?** (Children may realise that the amount of butter is double that of sugar in the original list, so he used 600 g of butter. Alternatively, using the table: 15 biscuits take 150g of sugar, 300 g = 2 × 150 g so double the amount of butter: 2 × 300 g = 600 g of butter.)

Unit 41: Identify 3-D shapes, including cubes and other cuboids, from 2-D representations

Content domain reference: 5G3b

Prerequisites for learning

Recognise 3-D shapes: cube, cuboid, cone, cylinder, other prisms and pyramids

Learning outcomes

Identify prisms from their 2-D nets

Identify pyramids and cones from their 2-D nets

Key vocabulary

Cube, cuboid, cone, cylinder, triangular, pentagonal, hexagonal, prism, triangular-based, square-based, pentagonal pyramid; 2-D representation, net, face

Resources

Resource 24: Nets of prisms (one per child and one enlarged to A3); Resource 25: Nets of pyramids and a cone (enlarged to A3); Resource 26: 3-D shapes and their nets; scissors; squared paper; set of 3-D shapes (that will open out to show their nets); squared paper

Background knowledge

A **prism** generally has two identical faces (one at at each end) and three or more flat faces that are rectangles or parallelograms. Cubes, cuboids and cylinders are all special types of prism. The total number of faces is two more than the number of sides on the end face. For example, a triangular prism has five faces.

A **pyramid** has a polygonal base and triangular faces. The number of triangular faces is equal to the number of sides of the base. The shape of the base gives the pyramid its name. A square-based pyramid has a square base and four identical triangular faces.

The net of a 3-D shape is a combination of 2-D shapes (the faces of the 3-D shape) that can be folded into the 3-D shape.

Teaching activity 1a (20 minutes)

Identify prisms from their 2-D nets

- Show children a set of 3-D shapes. Remind them of the definition of a prism. Ask: **Can you pick out all the prisms?** Explain that cubes, cuboids and cylinders are special types of prism because they all have two identical end faces.

- Children look at an enlarged copy of Resource 24: Nets of prisms and notice how all the nets of prisms are arranged. Say: **Each is made up of adjoining rectangles with an end shape on either side.** Give each child a copy of the resource and let them cut out and make one of the nets into a prism, then write the name of the prism on one of the faces.

- On squared paper, sketch a net of a different cuboid or triangular prism with the end shapes both on the **same** side of the rectangles. Ask: **Can we make this net into a shape?** Let a child try. Establish that although this diagram has the correct number of rectangular and end faces, it is not the net of a 3-D shape because it does not fold up correctly. Ask: **How can you tell if a diagram is the net of a prism? How many rectangular faces will a triangular prism have? How is the number of rectangular faces connected to the end shape? How many rectangular faces will the net of a hexagonal prism have? Does it matter where the end shape is on the net?** (*They must be on opposite sides of the rectangles, but it doesn't matter where.*)

Teaching activity 1b (20 minutes)

Identify prisms from their 2-D nets

- Use a set of shapes that open out into nets (or shapes made from Resource 24: Nets of prisms and Resource 25: Nets of pyramids and a cone). Select the prisms, including the cylinder, cubes and cuboids.

- Children each take one shape and carefully open it out to reveal its net. Look at each one in turn. Ask: **What was your shape before you unfolded it?** (*cuboid*) **This is called the net of the (cuboid) because it can be folded up into a (cuboid). What do you notice about each of these nets?** (*They are all made of rectangles and two other identical shapes.*) If possible (if the shapes can be taken apart), change the net, moving one or both of the end pieces along, keeping them on the same sides of the rectangles as they were. Ask: **Will this still fold up into a cuboid? Is it the same cuboid as before?** (*Yes*) Now move one of the end pieces so that both are on the same side of the rectangles and ask a child to refold it. Establish that the end pieces have to be on opposite sides of the rectangles.

- Describe what the net of a prism looks like. Ask: **Why is a cuboid [cube/cylinder] also a prism?** (*It is made from rectangles and has two identical end faces.*)

- Check children's understanding by asking them to sketch the nets of some prisms.

Teaching activity 2a (20 minutes)

Identify pyramids and cones from their 2-D nets

- Remind children of the definition of a pyramid. Show a set of 3-D shapes and ask them to pick out all the pyramids. Explain that a cone is not a pyramid because it does not have any triangular sides but, since it has other things in common (a base and a top point), the children are going to look at these shapes and their nets too.

- Look at how all the nets of pyramids are arranged on Resource 25: Nets of pyramids and a cone. Say: **The base shape has triangles on each of its sides. What do you notice about the triangles?** (*They are identical and isosceles.*) **Why is this one called a triangular pyramid?** (*It has a triangular base.*) **Why is this one called a square-based pyramid?** (*It has a square base.*) **What is the name of a pyramid that has a five-sided shape as its base?** (*pentagonal pyramid*) **What is the name of a pyramid with a hexagonal base?** (*hexagonal pyramid*)

- Children each cut out and make one of the shapes into a pyramid or cone and write the name on the base.

- Ask: **What do you think the net of a hexagonal pyramid would look like?** (*a hexagon with 6 triangles*)

- Cut the shapes from Resource 26: 3-D shapes and their nets and play a game, matching the shapes' nets to their names.

Teaching activity 2b (20 minutes)

Identify pyramids and cones from their 2-D nets

- Use a set of shapes that open out into nets (or shapes made from Resource 24: Nets of prisms and Resource 25: Nets of a pyramid and a cone). Separate out the prisms, including the cylinder, cube and cuboids from the other shapes.

- Children each take one of the remaining shapes and carefully open it out to reveal the net. Look at each net in turn, establishing that pyramids are made from a base shape with identical isosceles triangles. The number of triangles matches the number of sides on the base. The name of a pyramid comes from the shape of the base.

- Sketch some nets that look, initially, like pyramids, but with mistakes: triangles of different sizes, bases of triangles not matching the size of the base, triangular base with three squares or rectangles.

- Check children's understanding by asking them to sketch the net of a square-based pyramid, a hexagonal-based pyramid, a triangular pyramid.

- Cut the shapes from Resource 25: Nets of pyramids and a cone and play a game, matching the shapes' nets to their names.

Unit 42: Know angles are measured in degrees: estimate and compare acute, obtuse and reflex angles

Content domain reference: 5G4a

Prerequisites for learning

Identify acute and obtuse angles
Know that angles are a measure of turn

Learning outcomes

Identify acute, obtuse and reflex angles

Key vocabulary

Acute angle, obtuse angle, reflex angle; degree, °

Resources

Resource 27a: 2-D shapes (enlarged to A3, optional); squared or squared dotty paper; rectangles of tracing paper or card; rulers

Background knowledge

Acute angles are less than 90°, **obtuse** angles are between 90° and 180°, and **reflex** angles are between 180° and 360°. One right angle is a quarter turn (90°); two right angles produce a half-turn, which forms a straight line (2 × 90° = 180°); three right angles produce a three-quarter turn (3 × 90° = 270°); four right angles produce a full or whole turn (4 × 90° = 360°). The use of 360 is probably based on an old calendar in which there were 360 days in a year, the full revolution of the Earth around the Sun. It is a useful number because it can be divided exactly by 2, 3, 4, 5, 6, 8, 9, 10, 12, 15, 18, 20, 24, 30, 36, 40, 45, 60, 72, 80, 120 and 180.

Teaching activity 1a (15 minutes)

To identify acute, obtuse and reflex angles

- Ask children to remind you what angles are, the units they are measured in and the different types of angle that they know, classified by their size (acute, obtuse and right angle). Ask children to stand and face the front of the class. Then ask them to move a quarter turn to the right. Explain that they have turned through 90°. Repeat, so that they have turned a half-turn, 180°; repeat for a three-quarter turn, 270°; repeat to make full turn, 360°. Now they should be facing in the same direction they started in.

- Demonstrate on squared paper how to draw a right angle, using the grid lines. Show children how to mark the right angle with the correct square angle symbol. Write 'right angle, 90°, $\frac{1}{4}$ turn' underneath.

90°

- Repeat to show two right angles, 180°, $\frac{1}{2}$ turn, a straight line; then show three right angles, 270°, $\frac{3}{4}$ turn; and finally show four right angles, 360°, a whole turn, a full circle.

- Explain that all other angles lie between two multiples of 90°.
  - **Acute angles** are between 0° and a quarter turn: between 0° and 90°, but are smaller than 90°.
  - **Obtuse angles** are between a quarter and a half turn: more than 90° but less than 180°.
  - **Reflex angles** are between half and and a whole turn: more than 180° but less than 360°.

- Sketch some angles on squared paper for children to compare with your diagrams, marking each of them with the curved angle sign. The easiest way to draw a reflex angle is to draw an acute or obtuse angle, but label the outside.

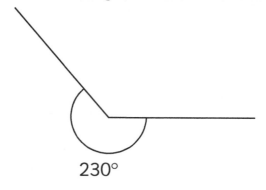

230°

- Point to some angles and ask the children questions: **Is it more than or less than 90°? Is it more than 90°? Is it more than 180°? Is it less than a right angle? Is it more than a right angle? Is it more than two right angles?**

- Children draw at least two examples of each type of angle: acute, obtuse and reflex. Ask them to label them correctly and then order them, from smallest to largest.

- For further practice, show children an enlarged copy of Resource 27a: 2-D shapes and ask them to name each of the angles within the shapes.

Teaching activity 1b (15 minutes)

To identify acute, obtuse and reflex angles

- Children revise angles, saying what they are, the units they are measured in and the different types of angle that they know, classified by their size (acute, obtuse and right angle).

- Ask children to show you a right angle with their arms – one straight up and the other out horizontally. Ask them to move their upright arm (and head) over towards their horizontal arm, making an acute angle. Ask: **How small an angle can you make?**

- Children return to the starting position for a right angle and repeat, moving the upright arm out wider to show an obtuse angle (but not a reflex angle).

- Ask them to show an angle of two right angles, 180°, with both arms horizontal and level. Then ask them to lower one arm to show a **reflex angle** – the angle between their arms, including their head.

- Draw an acute angle on a piece of squared paper, using the grid line for the base. Use a small curved line to show which angle you are measuring. Show how to use the corner and edges of tracing paper or card to show a right angle.

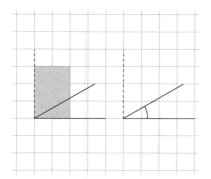

acute

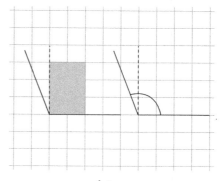

obtuse

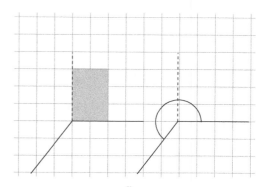

reflex

- Draw a dotted line along the paper edge to show 90° and then remove the paper. Point out that the angle you are measuring is less than 90° (1 right angle): **The line was under the paper, so it is less than 90°.**

- Repeat by drawing an obtuse angle, showing that the line is to the left (or right) of the 90° so is larger than 90°.

- Finally draw and mark a reflex angle. Show that this line is outside and below the paper guideline, so is more than 180°, so it is a reflex angle.

- Children draw two of each type of angle: acute, obtuse and reflex. Then they order them, from largest to smallest.

- For further practice, show children an enlarged copy of Resource 27a: 2-D shapes and ask them to name each of the angles within the shapes.

Unit 43: Draw given angles, and measure them in degrees
Content domain reference: 5G4c

Prerequisites for learning

Identify whether an angle is acute, obtuse or reflex
Know that angles are measured in degrees

Learning outcomes

Draw angles to the nearest two degrees
Measure angles to the nearest two degrees

Key vocabulary

Degree, protractor, acute, obtuse, reflex angle

Resources

Resource 27a: 2-D shapes; 180° protractors; large teaching protractor if available (or an online one); rulers; squared paper

Background knowledge

It is very important that children learn to use a protractor correctly. Common errors include positioning the protractor incorrectly, usually lining up the bottom edge of the protractor against the angle instead of the marked baseline and/or reading from the incorrect scale. Start by giving them angles with a horizontal baseline to measure, turning angles that are not given in this way so that one of the lines is horizontal. Children will usually find 180° protractors more manageable at this stage.

Teaching activity 1a (15 minutes)

Draw angles to the nearest two degrees

- Let children look closely at a protractor. Ask: **What does a protractor measure?** (*angles*) **What is the unit for measuring angles?** (*degrees*) **What is the largest angle this protractor measures?** (*180°*) **Why does it have two scales?** (*So you can measure angles from either side of the protractor.*) Explain that they are going to learn two ways of deciding which scale to use.

- Ask children to point to some numbers on the outer scale, starting with multiples of 10, then of 5, then any numbers. Repeat with the inner scale. Ensure that children realise that one scale reads left to right and the other right to left. Check that they are showing you the correct side of the appropriate multiple of ten.

- Demonstrate how to draw an angle of 40°. Draw a horizontal baseline on squared paper and mark a faint vertical dotted line at one end to show 90°. Place the protractor so that the middle half-moon shape is where the line and the dotted line meet and the line you have drawn is directly under the baseline on the protractor. Explain: **Because a degree is very small, we have to place the protractor very carefully to draw and measure angles accurately.**

Move the protractor about to show incorrect placing (half-moon is not on the exact end of the line; line is slanted; bottom edge of protractor is not against the drawn line). Move it back to the correct position. Say: **I want to draw an angle of 40°, which scale do I need to use? Which 40 is the correct one? Is 40° larger or smaller than 90°?** (*smaller*) Ask: **So which scale must it be?** Make a small pencil mark against the correct 40, take away the protractor and join the mark to the end of the line. Write 40° in the angle.

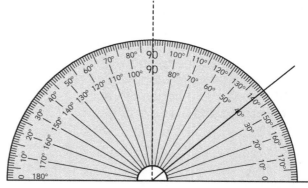

- Ask the children to try drawing angles of 60°, 45° and 32° .

- Repeat for 150°, establishing that 150 > 90. Say: **Choose the scale that will give a larger angle than the right angle we have already drawn.**

Teaching activity 1b (15 minutes)

Draw angles to the nearest two degrees

- Draw a line on squared paper and show children how to place the protractor correctly to draw the angle, explaining about the accuracy (see the diagram on the previous page). Say: **I am going to draw an angle of 40°, so I need to start at the scale that has a zero at the end of the line, so that I can find the correct 40. 40° is less than 90°, so it must be on this half of the protractor.** Indicate the 0 and follow the scale round with your finger, to 40. Make a mark. Complete the drawing by joining the mark to the end of the base line. Repeat, but draw the angle from the other end of the line, which will mean using the other scale.

- Repeat for 150°, drawing it from one end and then the other, showing that both scales are needed.

- As can be seen in the diagrams, the outer scale starts on the left, clockwise and the inner one on the right, anti-clockwise. To draw angles at the opposite ends of the line, you will need to use the inner scale for one and the outer scale for the other.

Teaching activity 2a (15 minutes)

Measure angles to the nearest two degrees

- Demonstrate how to measure some acute angles, then some obtuse angles, then alternate, using the strategy of comparing it with 90°. Draw a dotted line to show 90°. Then ask, for example: **Is the angle more or less than 90°? What sort of angle is it, acute or obtuse? So is it 125° or 55°?** Ensure that you draw some from the left of the line and some from the right. This strategy uses estimation by comparing with 90°. Identifying whether an angle is acute or obtuse will show which scale they need to use. Say, as appropriate: **It is obtuse, so it must be 125° not 55°. It is acute, so it must be 55° not 125°.**

- Measure some of the angles in the shapes on Resource 27a: 2-D shapes.

Teaching activity 2b (15 minutes)

Measure angles to the nearest two degrees

- Draw an acute angle from the left end of a line. Then demonstrate how to measure it, by correctly positioning the protractor, finding the correct scale on the right (the one that starts 0, not 180) and tracing your finger around until they can read the measurement.

- Draw a different acute angle from the right end of the line, using the correct scale on the left, which will be the other scale from the one you used previously. Repeat with two obtuse angles, one from the left of a line and one from the right, so that you use both the inner and outer scales.

- Measure some of the angles in the shapes on Resource 27a: 2-D shapes.

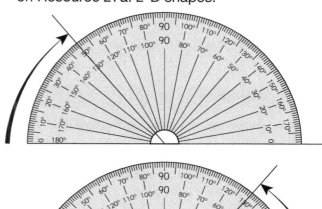

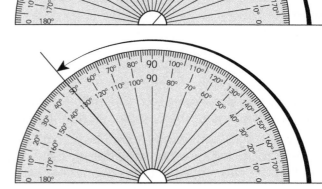

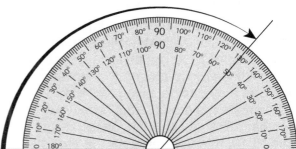

Unit 44: Identify: angles at a point and one whole turn (total 360°); angles at a point on a straight line and $\frac{1}{2}$ a turn (total 180°); other multiples of 90°

Content domain reference: 5G4b

Prerequisites for learning

Know that angles are a measure of turn and they are measured in degrees

Be able to read and draw angles accurately

Learning outcomes

Know that the sum of the angles at a point on a line is 180°

Know that the sum of the angles around a point is 360°

Key vocabulary

Half turn, whole turn, right angle, degrees

Resources

Tracing paper; 180° protractors; rulers; squared paper; prepared angles

Background knowledge

This lesson explores two important angle facts. The first is that the sum of the angles **at a point on a line** is 180°, which is also a half turn or two right angles. Children sometimes think that any angles on a line must add to 180° even if they do not start at the same point. The second is that the sum of the angles **around a point** is 360°, which is a whole turn or four right angles. These are both important angle facts used extensively in geometry in KS3 onwards.

Teaching activity 1a (15 minutes)

Know that the sum of the angles at a point on a line is 180°

- Ask children to draw a horizontal line on squared paper. Make sure they use rulers and sharp pencils. Then they mark a small point somewhere on the line and from it they draw another line slanting up left or right to form an angle either side of the point (not a right angle).

- They measure both angles carefully with a protractor, using their chosen strategy (Unit 43), writing the two angles in the relevant spaces.

- Repeat for several more pairs of angles.

- Then ask them to add up the two angles in each diagram. Ask: **What is the sum of these two angles?** Explain that the angles at a point on a line always add up to the same number. Can they decide, using rounding if their measurements are not accurate, what that number is? Give clues if necessary: **It is a multiple of 10; it is the same as the sum of two right angles.** Establish that the sum is 180°.

- Ask: **What do you notice about the type of angles on your lines? Are they both acute? Could they both be obtuse?** Establish that one is acute, one is obtuse: **Two obtuse angles would give a sum greater than 180°; two acute angles would give a sum less 180°.**

- Repeat, but now ask children to draw two lines from the point, giving three angles to measure and total. Ask: **Is the sum of the angles still 180°?** (*yes*) There can now be three acute angles, or two acute and one obtuse angle, or two acute and a right angle.

- Sketch to show two angles on a line, one marked 50°. Ask: **How can we work out what this other angle is? They must add to 180, so subtract 50 from 180 = 130.** Repeat with other pairs and then with three angles, two given and one to work out.

Teaching activity 1b (15 minutes)

Explore the sum of the angles at a point on a line

- Ask children to give you some pairs of multiples of 10 that add to 180. Note them down on slips of paper as number pairs to 180. Repeat with multiples of 5. Explain that the children are going to prove that the sum of the angles at a point on a line is 180°.

- Each child takes some tracing paper, squared paper, a protractor and a sharp pencil. They choose one of the slips of paper (number pairs to 180). Lay the tracing paper over the squared paper, draw a horizontal line, draw the acute angle from the right of the line (see diagram 1). Draw another line and draw the obtuse angle from the left of the line (see diagram 2). Cut them out roughly and join them together to make a line (see diagram 3). Repeat with another pair. Together, look at all the combined shapes they have made. Say: **We have proved that angles that add to 180° make a line. The angle fact is that the sum of the angles on a line is 180°.**

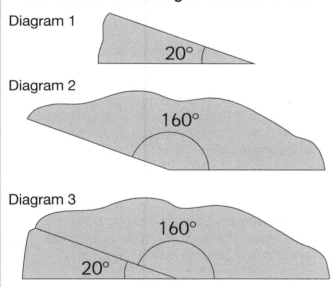

Diagram 1

20°

Diagram 2

160°

Diagram 3

160°

20°

- Draw a sketch to show two angles on a line, one marked 50°. Ask: **How can we work out what this other angle is?** *(The two angles add up to 180°, so subtract 50 from 180: 180° − 50° = 130°.)*

Teaching activity 2a (15 minutes)

Know that the sum of the angles around a point is 360°

- Ask children to draw two lines crossing each other, to give four angles around a point in the middle of the crossed line. They measure each angle carefully and work out the sum. They repeat this, then compare their results with those of other children to establish that the angle sum this time is 360°, a full turn. As a clue, draw

the two lines on grid paper to show four right angles: 4 × 90° = 360° (see diagram 4). There will be two pairs of angles: a pair of acute angles and a pair of obtuse angles. They should also discover that the opposite angles between the crossing lines are equal (see diagram 4).

- Draw two crossing lines, mark one angle and use that to work out the other angles, turning the paper so that children can see the two angles on a line. Repeat.

- Now mark a point on the paper and draw four lines radiating from the point. Measure the angles and work out the sum. Help children with any reflex angles. This sum should also be 360°. In diagram 5 the opposite angles are not equal, because there are four lines meeting at a point rather than two lines that cross.

- Sketch and mark all but one of the angles around a point for children to work out the missing angle. Repeat, marking one of the angles with the square right-angle symbol (see diagram 6).

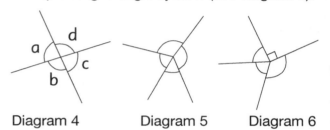

Diagram 4   Diagram 5   Diagram 6

Teaching activity 2b (15 minutes)

Explore the sum of the angles around a point

- Draw two sets of two angles from a point on a straight line, on tracing paper. Bring the pairs together, one pair on top and the other inverted underneath, as in diagram 7. Mark the point where all four angles meet. Ask: **What is the sum of the angles at the top? At the bottom?** *(These add to 180° and these also add to 180°.)* **What is the total of all four angles?** *(360°)* Say: **This shows that angles that add to 360° make a full turn. The angle fact is that the sum of the angles around a point is 360°.** The children could measure the angles to prove the angle fact.

- Sketch and mark all but one of the angles around a point, as shown in diagram 6, and ask children to work out the missing angle. Repeat, marking one of the angles with the square right-angle symbol.

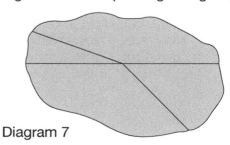

Diagram 7

Unit 45: Use the properties of rectangles to deduce related facts and find missing lengths and angles

Content domain reference: 5G2a

Prerequisites for learning

Find the area and perimeter of a rectangle

Key vocabulary

Rectangle, diagonal, bisect, parallel

Learning outcomes

Define and show diagrammatically the properties of a rectangle

Find missing angles and sides in rectangles

Resources

Squared paper; large 2-D rectangles; rulers; protractors; string

Background knowledge

Every flat (2-D) shape has **properties** relating to its sides and angles that define it and distinguish it from all other shapes. A rectangle has two pairs of opposite, equal, parallel sides and four equal angles (right angles). It has two diagonals which are equal in length. Where the diagonals cross, two pairs of angles are formed (opposite angles are equal) and the diagonals bisect each other (divide each other exactly into two equal halves).

Teaching activity 1a (15 minutes)

Define and show diagrammatically the properties of a rectangle

- Children draw three different rectangles on squared paper, ensuring that all the sides are on grid lines. They measure the angles. Ask: **Are all the angles the same size? What special type of angle are they? How can we show on our diagram that these are all right angles?** (*Use a square angle sign.*)

- Ask the children to measure each side, or count the squares, to see that opposite sides are equal. Show them how to mark the sides of a shape to show equal sides, explaining why each pair is marked in a different way. (Putting one marker on all of them implies that all the sides are equal, as shown in the diagram.)

- Discuss what 'parallel' means, using a ruler's edges to demonstrate. Say: **The distance between the lines will always be the same, however long the lines are.** Ask: **Are the opposite sides parallel?** Show how to use the ruler to measure the distance between the lines in several places, to show that they are parallel. Show how to mark parallel sides with arrows, then double arrows for a second pair of parallel sides, as in the diagram.

- Draw both diagonals on each rectangle. Measure them to see that they are the same length, then measure from the vertices to the centre to discover that they are all half the length of the diagonal, marking each half diagonal to show they are equal. Say: **The diagonals bisect each other, which means they cut each other in half.**

- Measure the angles at the centre, or recap how to show that opposite angles are equal (Unit 44). Demonstrate how to mark the equal angles. Measure the other two angles in the triangles formed by the diagonals, to show that the triangles are isosceles (two equal sides, two equal angles).

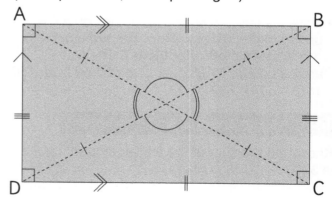

- Write the properties (special characteristics) of the rectangle as a list: two pairs of equal and parallel sides; four right angles, equal diagonals that bisect each other; two pairs of opposite, equal angles at the centre.

Teaching activity 1b (15 minutes)

Define and show diagrammatically the properties of a rectangle

- Show children 2-D rectangles of different sizes. Measure the angles, using a ruler or corner of a card to show that they are right angles. Measure the sides to show that the opposite sides are equal.

- Children draw round one of the rectangles and show these properties diagrammatically. Explain that having four right angles and two pairs of opposite and equal sides are two of the properties or special features of rectangles.

- Children each use a ruler on their rectangles to show that opposite sides are parallel.

- They use two pieces of string to measure the lengths of the diagonals, then compare them to show that they are the same size. They draw the diagonals onto the diagram.

- Help children to measure the angles at the centre or recap how to show that opposite angles are equal (Unit 44). Demonstrate marking equal angles. Measure the other two angles in the triangles formed by the diagonals to show that the triangles are isosceles.

- Recap the properties of rectangles, stressing that if just one of these properties is not true for a shape, then it is **not** a rectangle. A shape must have all of these properties to be a rectangle.

Teaching activity 2a (15 minutes)

Find missing angles and sides in rectangles

- Draw a rectangle and its diagonals on a whiteboard. Say: **We will use the properties of the rectangle to work out angles and sides without measuring.**

- Mark some different sides, lengths and angles and ask children to tell you or work out what the missing side or angle is.

  - Mark two adjacent sides 10 cm and 4 cm. Ask: **What is the perimeter of this rectangle?** (*28 cm*)

  - Mark one side 12 cm. Say: **The perimeter is 30 cm, what length is the missing side?** (*30 – (12 + 12) = 6; half of 6 is 3 cm*)

  - Mark the length of half a diagonal as 6 cm. Ask: **What is the length of the diagonal?** (*12 cm*)

  - Mark one of the acute angles at the centre as 70°. Ask: **What size are these other angles?** (*110°, 70°, 110°*)

  - Mark one of the angles in the isosceles triangle formed by the diagonals. Give one of the equal angles and ask for the size of the other one.

Teaching activity 2b (15 minutes)

Find missing angles and sides in rectangles

- Ask children to draw a rectangle on plain paper and mark one of the sides 10 cm. Ask: **Can we mark the length of any other side?** (*Opposite side is also 10 cm.*) Establish that we do not know the other sides yet. Say: **The perimeter is 36 cm, can we use that to find the length of the other two sides?** Point out that we know part of the perimeter is 10 + 10 = 20 cm. Ask: **What shall we do next?** (*36 – 20 = 16*) **Is that the answer? What do we need to do next?** (*Halve 16 because the other two sides are the same size.*)

- Ask children to draw another rectangle and mark one of the sides 12 cm. Say: **One side is 12 cm and the area of the rectangle is 36 cm^2.** Ask: **Can you use this information to find out the lengths of the missing sides?** (*The opposite side is 12 cm because it is the same. Then divide the area by one of the sides to find the other side: 36 ÷ 12 is 3. The missing sides are 3 cm.*)

- Draw another rectangle and mark the diagonals. Say: **This diagonal is 8 cm long. What is the length of the other diagonal?** (*8 cm*) Ask: **What is the length of this line (vertex to centre)?** (*4 cm, half of the diagonal*) **If this centre angle is 75°, what is the angle opposite it?** (*75°*) **What about this angle next to it?** (*180° – 75° = 105°*) Repeat with other rectangles.

- Draw another rectangle, mark the diagonals and ask questions related to the angles within the triangle formed by the diagonals. If one angle in the triangle is marked, it is possible to work out the other two, as they are isosceles, so the angles at the vertices are the same and the angle sum of a triangle is 180°.

Unit 46: Distinguish between regular and irregular polygons based on reasoning about equal sides and angles

Content domain reference: 5G2b

Prerequisites for learning

Know how to show equal angles and equal sides on shapes

Understand what is meant by the properties of a polygon

Learning outcomes

Define and name regular polygons with up to six sides

Use their properties to describe and identify polygons

Key vocabulary

Property, polygon, triangle (equilateral, isosceles, scalene, right-angled), rectangle, square, rhombus, parallelogram, kite, trapezium

Resources

Resource 27a: 2-D shapes; Resource 27b: 2-D shape properties; selection of 2-D shapes as shown on Resource 27a: 2-D shapes; squared paper; rulers; protractors; mini-whiteboards

Background knowledge

A polygon is, literally, a many-sided shape. Regular polygons must have all sides equal **and** all angles equal. Children often think that a shape is regular if it is angular or well known to them. They sometimes do not recognise pentagons and hexagons when they do not look like the shapes they know. The specific properties of a shape differentiate it from all other shapes, even if they share some of the properties.

Teaching activity 1a (10 minutes)

Define and name regular polygons with up to six sides

- Explain that for a shape to be called regular, it must have equal sides and equal angles. Children use the set of 2-D shapes (or Resource 27a: 2-D shapes laminated and cut out), measuring angles and sides where necessary, to sort them into regular and irregular polygons.

- When everyone is agreed, children draw a table with two columns, one headed 'regular polygons' and the other 'irregular polygons'. They sketch or draw round each shape under the correct heading and name the shape.

  ◆ Regular: equilateral triangle, square, regular pentagon and hexagon

  ◆ Irregular: all the other triangles and quadrilaterals. Draw an irregular pentagon and hexagon free hand.

Teaching activity 1b (15 minutes)

Define and name regular polygons with up to six sides

- Children use Resource 27a: 2-D shapes, a ruler and a protractor to identify the regular and irregular shapes by observation and measuring (where necessary). They mark any equal sides and angles (Unit 45). They write 'regular' or 'irregular' above each shape.

- Ask: **How do you know that a rhombus/trapezium/right-angled triangle/rectangle is not a regular shape?** (*They do not have all equal sides and equal angles.*) **Which is the only regular three-sided/four-sided polygon?** (*equilateral triangle/square*)

- Show a 'house-shaped' pentagon with equal sides. Ask: **What is the name of this shape?** (*pentagon*) **Is it a regular shape?** (*no*) **Why is this pentagon not a regular pentagon?** (*The angles are not equal.*) Ask children to draw a different irregular pentagon and an irregular hexagon.

Teaching activity 2a (15 minutes)

Use their properties to describe and identify polygons

- Explain that the properties of a shape describe its special features, so that it can be identified. Even if other shapes have some of the same properties, there will be at least one property that is different. Consider a shape and explain the special properties it has, such as parallel sides and right angles.

- Have two sets of 2-D shapes (or Resource 27a: 2-D shapes laminated and cut out) ready. Lay out one set and hide an identical one of the other set behind your back. Describe the shape, listing its properties one by one, asking after each statement: **What could my shape be and which shapes can we now discount?** For example, use these statements: **My shape has opposite sides that are equal. It is not a regular shape. Which shapes can we now discount? It does not have any right angles, but opposite angles are equal. Not all its sides are equal. What is my shape?** Children draw or write parallelogram or pick it out from the set of shapes. Repeat with a different shape

- Now play 'Ten questions'. Again, choose and hide a shape. Children must ask questions that involve only one property, for example, opposites that are equal, and to which the answer must be 'yes' or 'no'. They must not guess the shape until they are sure they know which shape it is.

- Repeat with a child choosing and hiding a shape and answering 'yes' or 'no'.

- If not all shapes are available, the shapes could be drawn for others to guess.

- To check understanding, cut out the cards on Resource 27b: 2-D shape properties and ask children to choose a card and read the properties, then another child picks out or identifies the correct shape.

Teaching activity 2b (15 minutes)

Use their properties to describe and identify polygons

- Remind children who completed activity 1b that they have shown some of the properties of the polygons by marking the equal sides and angles. Discuss some of the other properties, such as right angles and parallel sides, marking them accordingly (Unit 45).

- Use cards cut from Resource 27a: 2-D shapes and Resource 27b: 2-D shape properties to play a matching game. Lay them all out, face down, keeping them in two sets: 'shapes' and 'properties'. Children take turns to pick one from each set: if they get a match they keep the cards, if not they return them to where they came from and the next player has a turn. They continue until the cards are all paired up. An easier, quicker version would be to see how quickly, in pairs or groups of four, they can match up each shape with its list of properties.

- Separate out the properties cards and read some of them out, one at a time. Children draw the correct shape on their mini-whiteboards, showing any equal sides or angles and right angles in the correct way.

Unit 47: Identify, describe and represent the position of a shape following a reflection or translation, using the appropriate language, and know that the shape has not changed

Content domain reference: 5P2

Prerequisites for learning

Use the language of up, down, right, left to describe movements between positions

Read and plot coordinates in the first quadrant

Learning outcomes

Describe and translate shapes in a first quadrant coordinate grid

Describe and reflect shapes in a first quadrant coordinate grid

Key vocabulary

Reflection, translation, reflect, translate, move, position, quadrilateral, left, right, up, down

Resources

Resource 28: Coordinate grids; 2-D shapes: triangles and quadrilaterals; tracing paper; mirrors

Background knowledge

Reflections and translations are called **transformations** in mathematics.

When an object is **reflected**, it produces a mirror image on the opposite side of the line of reflection. The mirror line is called the axis of reflection. The size and shape of the object does not change, however the orientation of the shape (the way it is facing) generally does change.

When an object is **translated**, it moves to a different position. The size, shape and orientation of the object do not change. Translations were introduced in Year 4. They can also be described in terms of a vector in the form $\binom{x}{y}$ where x is the movement left–right and y is the movement up–down, so provide children with plenty of practice moving left–right first, then up–down.

At this stage, when drawing a reflected or translated image on a coordinate grid, any sides of the original shape that are on grid lines should also be on grid lines in the image. This is one method for spotting errors.

In the first quadrant, all coordinates are positive.

Teaching activity 1a (20 minutes)

Describe and translate shapes in a first quadrant coordinate grid

- Use Resource 28: Coordinate grids to revise how to read and plot coordinates in the first quadrant. Plot a point on the grid and let children read it. Then say a point and let children plot it. Give some reminders of how to remember which comes first: **Along the corridor, up the stairs; x is a cross (across).** Ask children: **Can you can remember what a translation is?** Explain that a translation moves a shape to the left or right, up or down, or a both. Place a shape on a coordinate grid, as in Resource 28: Coordinate grids, and draw around it. Keeping all movements to whole numbers of squares, move it up and then

back to its original position, move it right and back, down and back, left and back. Now move it up and to the left, down and to the right. Explain: **These all show movements or translations of the shape. The shape has moved but its shape, size and orientation (the way it is facing) have not been changed.**

- Fit the shape so that the top edge or vertex is directly on a grid line, or on a point where grid lines cross. Mark where one vertex (top or top left) is on the grid with a cross. Move the shape in one direction only. Ask: **Who can describe how the shape has been moved?** (*2 squares to the right, 4 squares up*) Mark a cross on the grid to show the new position of the same vertex. Remove the shape and check to see that the position of the cross has changed as described. Repeat for each of the four directions, taking it back to the starting position each time and marking the new position and counting the squares between the starting crosses and the new position of the cross, to check.

- Place the shape on a clean grid and again mark the position of one vertex. Move the shape in two directions: left or right first, then up or down. Ask children to describe the movement, giving left/right first (*2 squares to the left and three squares down; right four, up six*). Again, mark the position of the chosen vertex in each new position and count the squares right/left then up/down to check.

- Make a note of how the coordinates have changed.

- Give a translation (right/left movement first: for example, two right, three up) and ask a child to move the shape to its new position. Repeat several times and make a note of how the coordinates have changed. Ask: **How is the change in coordinates linked to the numbers in the translation?** (*A move to the right increases the x-coordinate by that amount, to the left decreases it; a movement up will increase the y-coordinate by that amount and a movement down will decrease it.*)

- Repeat with a different shape, remembering to mark the starting and finishing positions of one of the vertices. Children first describe a translation, then move the shape according to the instruction.

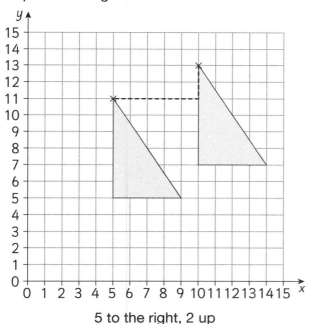

5 to the right, 2 up

Teaching activity 1b (20 minutes)

Describe and translate shapes in a first quadrant coordinate grid

- Use Resource 28: Coordinate grids to revise how to read and plot coordinates in the first quadrant. Plot a point on the grid and let children read it. Then say a point and let children plot it. Give some reminders of how to remember which comes first: **Along the corridor, up the stairs; x is a cross (across).** Ask the children what they remember about translation from Year 4. Demonstrate, with a shape on a coordinate grid. Explain that a translation moves a shape to the left or right, up or down, or a both. Move the shape back to its original position, move it right and back, down and back, left and back. Now move it up and to the left, down and to the right. Explain: **These all show movements or translations of the shape. The shape has moved but its shape, size and orientation (the way it is facing) have not been changed.**

- Draw a simple 2-D shape on Resource 28: Coordinate grids, ensuring that every vertex is on a point where grid lines cross. Label the shape A. Draw exactly the **same** shape in a different position on the grid, again with every vertex on a point where grid lines cross, and label it B. Explain that shape A has been translated or moved to a new position on the grid, shape B. Ask: **Who can describe how it has been moved?** Show how to mark the vertices, before and after the translation, and count squares to find out how **each** vertex has moved. Show that all the vertices have moved in the same way, for example, 3 left, 4 down. Ask: **What movement will translate B back to A?** (*3 right, 4 up, or the opposite translation to the first*)

- Repeat, drawing the same shape in another position and labelling it C. Ask: **What movement has translated A to C? Does it work for each vertex on the shape?**

- Discuss what has happened to the coordinates and how this is linked to the translation. Moving right/up increases the *x/y* coordinate, moving left/down decreases the *x/y* coordinate.

- Now ask children to describe the movement from C to A, from C to B and from B to C. Ask: **What do you notice about the translation A to C and C back to A?** (*They use the same numbers, but left becomes right, right becomes left, up becomes down, down becomes up.*)

- On a new grid, draw a different shape, with vertices all on points where grid lines cross, as before. Demonstrate how to translate a shape to the given instruction: **Six to the right, five down**:
  - Start at one vertex, count six squares to the right then seven down. Put a cross on the relevant coordinate point.
  - Repeat with each vertex in turn until you have translated them all. Join up the crosses into the shape that looks like the one you started with, but in a different position. If the shape is complicated, it is better to join up the crosses as you go, as it may be difficult to see the shape if there are too many vertices/crosses.

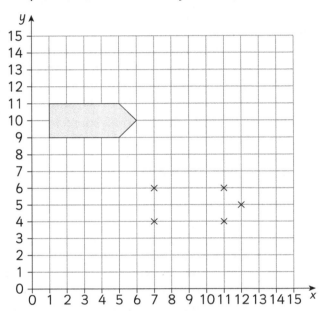

- Repeat with a different shape and a new translation.

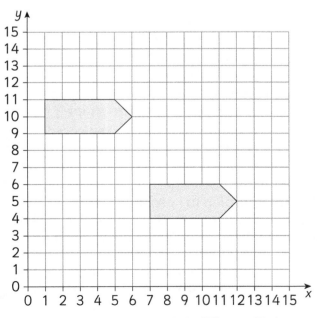

- Discuss the coordinates. Ask: **What will the new coordinate be when this shape has been translated 5 to the right? 4 to the left and 2 down?**

Teaching activity 2a (20 minutes)

Describe and reflect shapes in a first quadrant coordinate grid

- Use Resource 28: Coordinate grids to revise how to read and plot coordinates in the first quadrant.
- Ask children to describe a reflection (possibly as an image in a mirror or shiny surface). Explain that there are also reflections in maths. Say: **A reflection changes or transforms a shape in a special way.**
- Demonstrate how to use tracing paper to reflect a shape in a vertical line.
  - Draw a right-angled triangle on the right side of the grid on Resource 28: Coordinate grids, ensuring that all the vertices are on points where grid lines cross.
  - Draw a vertical line of reflection three squares to the left of the shape.
  - Place the tracing paper over the shape and reflection line, so that its edge is in line with the grid (landscape or portrait).
  - Trace the shape **and** the reflection line, making a small dot or cross on the traced reflection line and on the grid underneath.
  - Flip the tracing paper over, match up the reflection line and the dot or cross.
  - Trace hard over the reflected line to show where on the grid the reflection is.
  - Remove the tracing paper and draw the reflected shape in the correct position.

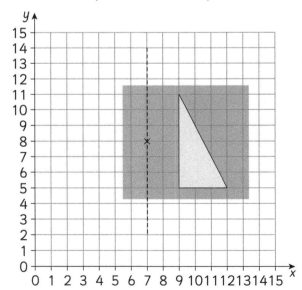

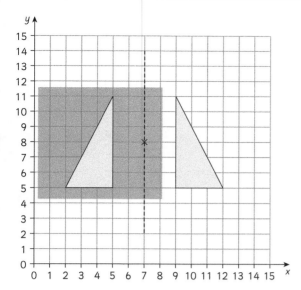

- Read the coordinates of both the original shape and of its reflection. Ask: **Which coordinate has changed?** (the *x*-coordinate) **Which coordinate has stayed the same?** (the *y*-coordinate)

- Repeat with a different shape and a horizontal reflection line with the object below the line of reflection. Ask: **Which coordinate has changed?** (the *y*-coordinate) **Which coordinate has stayed the same?** (the *x*-coordinate).

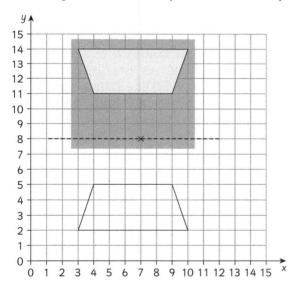

Teaching activity 2b (20 minutes)

Describe and reflect shapes in a first quadrant coordinate grid

- Use Resource 28: Coordinate grids to revise reading and plotting coordinates in the first quadrant. Draw a right-angled triangle on the right side of a coordinate grid on Resource 28: Coordinate grids, aligning as many edges to the grid lines as possible (preferably all of them). Leave it on the grid.

  - Draw a vertical reflection line three squares to the left of the shape.

- Place a mirror on the line of reflection to see where the triangle should move to. Flip the triangle over vertically and move it to where the image in the mirror was, leaving the outline you have drawn. Count the squares from the line of reflection to the outline showing the original position and to the new position of the triangle. Ask: **Are they the same distance from the line of reflection?** Adjust so that the shape is in exactly the right position, the same distance from the mirror line and lined up with the intersections of the grid lines. Draw carefully around the shape in its reflected position.

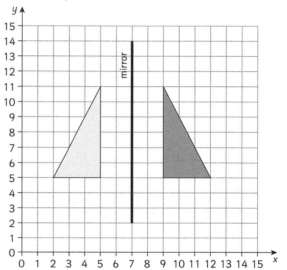

- Read the coordinates of both the original shape and of its reflection. Ask: **Which coordinate has changed?** (the *x*-coordinate) **Which coordinate has stayed the same?** (the *y*-coordinate)

- Repeat with a different shape and a horizontal reflection line with the object below the line of reflection. Flip the shape over horizontally. Ask: **Which coordinate has changed?** (the *y*-coordinate) **Which coordinate has stayed the same?** (the *x*-coordinate).

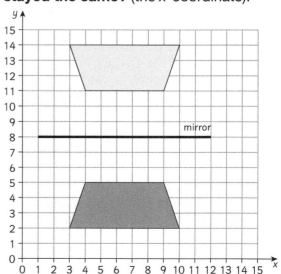

Unit 48: Solve comparison, sum and difference problems using information presented in a line graph

Content domain reference: 5S2

Prerequisites for learning

Understand the intervals in graph scales

Key vocabulary

Line graph, scale, interval

Learning outcomes

Read and interpret conversion lines graphs
Read and interpret other continuous data lines graphs

Resources

Resource 29: Line graphs; Resource 30: Line graph template; rulers

Background knowledge

A line graph is made up of points connected by lines. It shows how something changes in value. The parts of the line between the points usually have meaning. For example, conversion graphs between metric and imperial units, or between different currencies, produce a straight line sloping up: as one value increases, so does the other value. Other line graphs show changes in a variable, such as temperature, over time. The line may go up and/or down depending on whether the value of the variable increases or decreases.

Teaching activity 1a (20 minutes)

Read and interpret conversion line graphs

- Use Resource 29: Line graphs to discuss in detail how a line graph works. Look at the first graph, the conversion graph. Ask: **What does this line graph show? What does it do?** (*It changes between gallons and litres.*) **Which part shows gallons? Is the scale the same for both? What is each square worth on the litre scale?** (*1 litre*) **What is each square worth on the gallons scale?** (*half a gallon*) Explain that the scales do not need to be the same on both axes. The scale is decided by how much data needs to be shown.

- Show how to use a ruler to connect 2 gallons to the litre scale, as shown by the dotted lines: place the ruler to point vertically up from the 2 on the gallons scale. Draw a line up to where it crosses the graph line. Place the ruler horizontally across the graph, from the place where the vertical line meets the graph line, across to the litres line, draw a line and read the value on the litre scale: 9 litres. Write 2 gallons = 9 litres.

- Repeat for 4 gallons, then 5 gallons, then $3\frac{1}{2}$ gallons. Show how to find these values on the gallons scale and to interpret the results to the nearest litre.

- Use a similar method to show how to connect the litre scale to the gallons scale, first drawing a horizontal line from the litres axis across to the graph line, then another line down to the gallons axis. Start with 20 litres, then use values between the marked intervals, showing how to interpret the results to the nearest half or even quarter gallon.

- Explain that all the points on the line have meaning. Choose a point on the line between intervals and work out what it shows, for example, 2.5 gallons is approximately 11 litres.

Teaching activity 1b (20 minutes)

Read and interpret conversion line graphs

- Explain that line graphs are used to show the connection between two sets of data, for example, changing between currencies or between metric and imperial units. Write on a whiteboard '4.5 litres = 1 gallon'. Work out four more equivalents: 9 litres = 2 gallons, 4 gallons = 18 litres, 8 gallons = 36 litres, 6 gallons = 27 litres.

- Explain that you are going to draw a line graph to show this information. Use Resource 30: Line graph template and write the title 'Conversion graph for gallons and litres'.

- Ask: **Where shall I write gallons? Where shall I write litres? How many litres/gallons do I need to show on my graph?** Discuss an appropriate scale. Say: **There are 15 squares up and 10 across. We need to show at least 8 gallons, so one square per gallon works. We need to show at least 36 litres, so can we use one square per litre?** (no). **Two litres per square?** (*not quite*) Discuss the idea that 3 is not a very helpful number as it would be very difficult to plot 4.5 litres on that scale, but 4s would work well.

- Label the axes: Gallons, scale 1, 2, 3, 4 …10 on the horizontal axis and Litres, scale 4, 8… 48 on the vertical axis. There is no need to continue further. Show children how to use the pairs of equivalent values to plot four points, using two rulers, one up from the gallons and one across from the litres. If they are plotted accurately the points will join up into a straight line. Continue the line down to the origin (0, 0) and up beyond the last point.

- Use the graph, as in activity 1a, to read some values.

Teaching activity 2a (20 minutes)

Read and interpret other continuous data line graphs

- Show children the second graph (heart rate) on Resource 29: Line graphs and ask: **What does this line graph show? How is it different from the conversion graph?** (*The line is not straight, it goes up and down.*) **What do the scales show?** (*heartbeats per minute and time in minutes*) **What is each square worth on the minutes/heartbeat scale?** (*10*)

- Discuss why the line changes over time: it shows the number of heartbeats per minute before, during and after exercise. Ask: **What happens to the heartbeat when someone exercises?** (*It increases.*) **Will it keep on increasing?** (*No, once it reaches a certain level it will stay the same.*) **When does it decrease again?** (*When the person slows down or stops exercising.*)

- Ask: **Why does the line not start or end at zero?** If necessary, explain that the heartbeat of a living person is never 0. Say: **Where the line starts is called the resting heart rate.** Ask: **What is the resting heart rate of this person?**

- Ask: **What is the maximum heart rate shown here?** (*70*) **How long is the period of exercise?** (*30 minutes, from where the line goes up to when it comes down*) **How long was the heart rate at is maximum?** (*20 minutes, shown by the straight horizontal line*)

Teaching activity 2b (20 minutes)

Read and interpret other continuous data lines graphs

- Remind children that line graphs are used to show the connection between two variables. If possible, conduct an experiment to show that heart rate increases during exercise. Let them use their wrist or neck pulse to count their heart rate for 15 seconds, then multiply it by 4 to work out their resting heart rate. Working in pairs with similar resting rates, children jog around the playground for 5 minutes. Then one takes their own pulse again and records it while the other continues jogging for another 5 minutes. The second child takes their pulse after 10 minutes. Then they walk back to the classroom and take their pulse again. If it is not back to resting rate they take it again a few minutes later.

- Use Resource 30: Line graph template to record the results. Choose a set of results that represents what really happens (rest rate between 70 and 80, up to 110–130, remaining at maximum for 10–20 minutes, then 5–15 minutes to return to resting rate) or use similar values to the second graph on Resource 29: Line graphs. (Children's results may be inaccurate as they often find it difficult to feel their pulse, so count unreliably.) Show how to plot each section, then use a ruler to join up the points with straight lines.

- Talk about the graph they have created. Ask: **Why does the line not start or end at zero?** (*If the heartbeat is 0 the person is not alive.*) Say: *Where the line starts is called the resting heart rate.* Ask: *What is the resting heart rate of this person?*

- Ask: **What is the maximum heart rate shown here? How long is the period of exercise? How long was the heart rate at is maximum?**

Unit 49: Complete, read and interpret information in tables, including timetables

Content domain reference: 5S1

Prerequisites for learning

Add and subtract two-digit numbers

Work out time intervals given in 24-hour digital format

Learning outcomes

Read and interpret timetables

Read and interpret two-way tables

Key vocabulary

Timetable, two-way table, data, survey

Resources

Resource 31: Tables templates (prepared for 1a and 2a); Resource 32: Two-way table templates; local leaflets displaying opening times; bus timetables; geared clocks (optional)

Background knowledge

Typical transport timetables use 24-hour digital times. Children often forget that there are only 60 minutes in an hour, not 100, when working out durations that cross a whole hour. This means that simply using a column method to add or subtract will generally not work. There are a wide variety of real-life tables used and on display on notice boards and in leaflets. Local leaflets with tables are a useful and free resource and help children to see real-life applications.

Teaching activity 1a (15 minutes)

To read and interpret timetables
- Prepare Resource 31: Tables templates by making up or copying part of a local timetable into the first table, in a similar way to Question 1 on the practice page. You could manipulate the times so that the journey times from first to last stop are the same (unless you want to compare longer and shorter journeys). Ensure that some parts of the journeys go over an hour boundary. Use 24-hour times, with just one time at each stop, rather than arrival and departure times.
- You could use a real timetable, but they are often quite complicated as they cover a whole day and often journey times, as well as times between buses, vary. Isolate part of the timetable for children to look at.
- Look together at the timetable, pointing out how the times are written. Ask some key questions: **How long does it take to get between A and B (not always between the first and last stops)? If Allie wants to get to B by 11:30, what time is the latest bus she can catch? Pavel takes 10 minutes to walk to the bus-stop, what time will he have to set off to catch the 11:06 bus? The 10:30 bus is running 12 minutes late, what time will it arrive at C? Jaz arrives at her bus-stop in D at 12:59, how long will she have to wait for her bus? How frequently do the buses run between (for example) 10:00 and 12:00?** Repeat with a timetable showing opening times.

Teaching activity 1b (15 minutes)

To read and interpret timetables
- Show as many leaflets with timetables as you can, explaining how they work.
- Give each child a copy of Resource 31: Tables templates. Ask them, together, to make up five names of places to write in the first column.
- Ask a child to suggest a time between 10 and 10:30. They write that time against the first 'stop'. Say: **The bus takes 12 minutes to get to the next stop, what time will it arrive?** Use geared clocks to help, where necessary. Write in the agreed time. Repeat with different times, ensuring that at least one interval will need to go over into the next hour. Ask some questions relating to time intervals between non-consecutive stops.
- Explain that bus timetables often give the same journey time between bus stops. Ask for another start time, between 11:30 and 12:00, writing it in the second column. Children then have to work out the time interval between consecutive stops, using the times in the first column, and apply those to the new starting time.
- Repeat, ensuring that children start with a time after 1 o'clock and write it in 24-hour digital time.
- Ask questions, as in Activity 1a, including some of the form: **Which bus does she need to catch to get to X by a certain time?** and some relating to buses being early and late.

Teaching activity 2a (15 minutes)

Read and interpret two-way tables

- Prepare Resource 32: Two-way table templates. Ideas could include: boy or girl is left-handed or right-handed, wearing glasses, does not wear glasses, is in Year 3, Year 4, Year 5, has more than one sibling, fewer than one sibling, born in a specific town, or not. Do not write any totals.

- Explain that this table shows the data (information) from a survey. Discuss how the two-way table works by asking questions relevant to the labels: **How many left-handed girls are there in the survey? How many boys wear glasses? How many boys are there in Years 4 to 6?**

- Ask questions that will need a little more work: **How many girls were surveyed altogether? How many people were surveyed?** Fill in the totals.

- Show another two-way table with only some of the values written in. Ask: **Can we fill in any of the other values without any further information?** Provide the information they need, one bit of information at a time: **There are 28 children in Year 5.** (*Complete Year 5: 28 – 5 = 23.*) **A quarter of children in Year 4 are left-handed.** (*Complete Year 4: 24 is three-quarters so one quarter is 8; 24 + 8 = 32.*) Ask: **Do we need any more information?** Establish that as you now know the totals for Year 4 and Year 5, you can use this to work out the total for Year 6 by subtracting from 84, then use this to work out how many left-handed pupils there are in Year 6 (*3*).

	Year 4	Year 5	Year 6	Total
Right-handed	24		21	
Left-handed		5		
				84

Teaching activity 2b (15 minutes)

Read and interpret two-way tables

- Start with an empty two-way table, filling in the labels, as below or ask children to suggest what the labels could be. Explain that the the lables must be 'either or': if a person is in one category, they cannot be in any other category. You are either left-handed or right-handed, you are in Year 4 or in Year 5, you have siblings or you don't. Make some statements relevant to the categories, asking children to fill in what you tell them (see **bold** numbers in the table).

  - **There are 26 children in Year 4, 7 of whom wear glasses.**
  - **A third of the 30 children in Year 6 wear glasses.**
  - **57 children do not wear glasses.**
  - **82 children were surveyed.**

- They can then work out the missing values [in brackets] with what they have:
82 – 57 = 25 wear glasses; 25 – 7 – 10 = 8 in Year 5 wear glasses; 82 – (26 + 30) = 26 in Year 5.

	Year 4	Year 5	Year 6	Total
Wear glasses	7	[8]	10	[25]
Not wear glasses	[19]	[18]	[20]	57
	26	[26]	30	82

Place-value grids

millions	hundred thousands	ten thousands	thousands	hundreds	tens	ones			
							numerals		
								digit values	
									words

millions	hundred thousands	ten thousands	thousands	hundreds	tens	ones			
							numerals		
								digit values	
									words

millions	hundred thousands	ten thousands	thousands	hundreds	tens	ones			
							numerals		
								digit values	
									words

millions	hundred thousands	ten thousands	thousands	hundreds	tens	ones			
							numerals		
								digit values	
									words

millions	hundred thousands	ten thousands	thousands	hundreds	tens	ones			
							numerals		
								digit values	
									words

Negative number lines

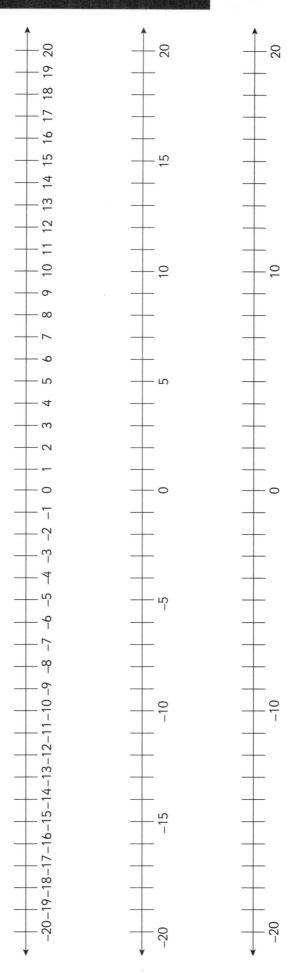

Thermometers

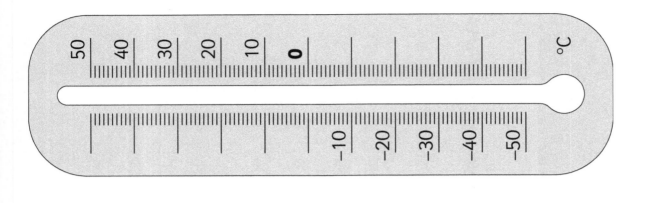

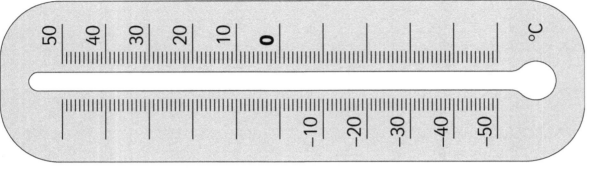

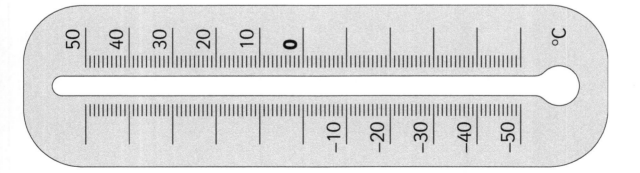

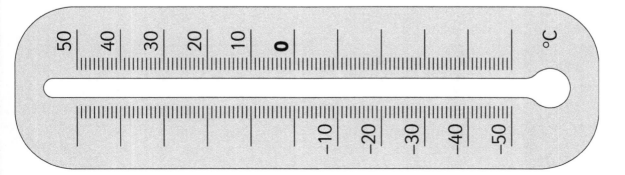

Roman numerals 1–100

The rules

1 Roman numerals are written in order of place value, starting with the largest.
CCLXVIII is 100 + 100 + 50 + 10 + 5 + 3 = 268.

2 When a symbol with a smaller value is written before a symbol with a larger value, subtract it. These all involve the decimal digit 4 or 9, but there are other rules that govern what you can and cannot write in this way.

- I can be subtracted from V and X but not from L, C, D or M.
- IV means 5 – 1 = 4, IX means 10 – 1 = 9
- X can be subtracted from L and C but not from D or M.
- XL means 50 – 10 = 40, XC means 100 – 10 = 90
- C can be subtracted from D or M.
- CD means 500 – 100 = 400, CM means 1000 – 100 = 900

3 It is never necessary to repeat the symbol V, L or D. A double gives a total that has its own letter.
VV is 5 + 5 = 10 (X), LL is 50 + 50 = 100 (C), DD is 500 + 500 = 1000 (M).

4 The symbols I, X, C and M can be used a maximum of **3** times:
II means 2, XXX means 30, CC means 200, MMM means 300.

1	I	21	XX1	41	XLI	61	LXI	81	LXXXI
2	II	22	XXII	42	XLII	62	LXII	82	LXXXII
3	III	23	XXIII	43	XLIII	63	LXIII	83	LXXXIII
4	IV	24	XXIV	44	XLIV	64	LXIV	84	LXXXIV
5	V	25	XXV	45	XLV	65	LXV	85	LXXXV
6	VI	26	XXVI	46	XLVI	66	LXVI	86	LXXXVI
7	VII	27	XXVII	47	XLVII	67	LXVII	87	LXXXVII
8	VIII	28	XXVIII	48	XLVIII	68	XLVIII	88	LXXXVIII
9	IX	29	XXIX	49	XLIX	69	LXIX	89	LXXXIX
10	X	30	XXX	50	L	70	LXX	90	XC
11	XI	31	XXXI	51	LI	71	LXXI	91	XCI
12	XII	32	XXXII	51	LII	72	LXXII	92	XCII
13	XIII	33	XXXIII	53	LIII	73	LXXIII	93	XCIII
14	XIV	34	XXXIV	54	LIV	74	LXXIV	94	XCIV
15	XV	35	XXXV	55	LV	75	LXXV	95	XCV
16	XVI	36	XXXVI	56	LVI	76	LXXVI	96	XCVI
17	XVII	37	XXXVII	57	LVII	77	LXXVII	97	XCVII
18	XVIII	38	XXXVIII	58	LVIII	78	LXXVIII	98	XCVIII
19	XIX	39	XXXIX	59	LIX	79	LXXIX	99	XCIX
20	XX	40	XL	60	LX	80	LXXX	100	C

Roman numeral cards

I	I	I
X	X	X
C	C	C
M	M	M
V	L	D

1	10	1000
1	100	1000
1	100	5
10	100	50
10	1000	500

100 square

1	2	3	4	5	6	7	8	9	10
11	12	13	14	15	16	17	18	19	20
21	22	23	24	25	26	27	28	29	30
31	32	33	34	35	36	37	38	39	40
41	42	43	44	45	46	47	48	49	50
51	52	53	54	55	56	57	58	59	60
61	62	63	64	65	66	67	68	69	70
71	72	73	74	75	76	77	78	79	80
81	82	83	84	85	86	87	88	89	90
91	92	93	94	95	96	97	98	99	100

Multiplication grid

Multiplication square: 1–12 times tables

×	1	2	3	4	5	6	7	8	9	10	11	12
1	1	2	3	4	5	6	7	8	9	10	11	12
2	2	4	6	8	10	12	14	16	18	20	22	24
3	3	6	9	12	15	18	21	24	27	30	33	36
4	4	8	12	16	20	24	28	32	36	40	44	48
5	5	10	15	20	25	30	35	40	45	50	55	60
6	6	12	18	24	30	36	42	48	54	60	66	72
7	7	14	21	28	35	42	49	56	63	70	77	84
8	8	16	24	32	40	48	56	64	72	80	88	96
9	9	18	27	36	45	54	63	72	81	90	99	108
10	10	20	30	40	50	60	70	80	90	100	110	120
11	11	22	33	44	55	66	77	88	99	110	121	132
12	12	24	36	48	60	72	84	96	108	120	132	144

Prime and composite numbers completed

N	Factors	Product	N	Factors	Product
1			41	1, 41	Prime
2	1, **2**	Prime	42	1, **2, 3**, 6, 7, 14, 21, 42	2 × 3 × 7
3	1, **3**	Prime	43	1, **43**	Prime
4	1, **2**, 4	2 × 2	44	1, **2**, 4, **11**, 22, 44	2 × 2 × 11
5	1, **5**	Prime	45	1, **3, 5**, 9, 15, 45	3 × 3 × 5
6	1, **2, 3**, 6	2 × 3	46	1, **2, 23**, 46	2 × 23
7	1, **7**	Prime	47	1, 47	Prime
8	1, **2**, 4, 8	2 × 2 × 2	48	1, 2, 3, 4, 6, 8, 12, 16, 24, 48	2 × 2 × 2 × 2 × 3
9	1, **3**, 9	3 × 3	49	1, 7, 49	7 × 7
10	1, **2, 5**, 10	2 × 5	50	1, 2, 5, 10, 25, 50	2 × 5 × 5
11	1, 11	Prime	51	1, 3, 17, 51	3 × 17
12	1, **2, 3**, 4, 6, 12	2 × 2 × 3	52	1, 2, 4, 13, 26, 52	2 × 2 × 13
13	1, **13**	Prime	53	1, 53	Prime
14	1, **2, 7**, 14	2 × 7	54	1, 2, 3, 6, 9, 18, 27, 54	2 × 3 × 3 × 3
15	1, **3, 5**, 15	3 × 5	55	1, 5, 11, 55	5 × 11
16	1, **2**, 4, 8, 16	2 × 2 × 2 × 2	56	1, 2, 4, 7, 8, 14, 28, 56	2 × 2 × 2 × 7
17	1, **17**	Prime	57	1, 3, 19, 57	3 × 19
18	1, **2, 3**, 6, 9, 18	2 × 3 × 3	58	1, 2, 29, 58	2 × 29
19	1, 19	Prime	59	1, 59	Prime
20	1, **2**, 4, 5, 10, 20	2 × 2 × 5	60	1, 2, 3, 4, 5, 6, 19, 12…	2 × 2 × 3 × 5,
21	1, **3, 7**, 21	3 × 7	61	1, 61	Prime
22	1, **2, 11**, 22	2 × 11	62	1, 2, 31, 62	2 × 31
23	1, **23**	Prime	63	1, 3, 7, 9, 21, 63	3 × 3 × 7
24	1, **2, 3**, 4, 6, 8, 12, 24	2 × 2 × 2 × 3	64	1, 2, 4, 8, 16, 32, 64	2 × 2 × 2 × 2 × 2 × 2
25	1, **5**, 25	5 × 5	65	1, 5, 13, 65	5 × 13
26	1, **2, 13**, 26	2 × 13	66	1, 2, 3, 11, 22, 33, 66	2 × 3 × 11
27	1, **3**, 9, 27	3 × 3 × 3	67	1, 67	Prime
28	1, **2**, 4, 7, 14, 28	2 × 2 × 7	68	1, 2, 4, 17, 34, 68	2 × 2 × 17
29	1, **29**	Prime	69	1, 3, 23, 69	3 × 23
30	1, **2, 3, 5**, 6, 10, 15, 30	2 × 3 × 5	70	1, 2, 5, 7, 10, 14, 35, 70	2 × 5 × 7
31	1, 31	Prime	71	1, 72	Prime
32	1, **2**, 4, 8, 16, 32	2 × 2 × 2 × 2 × 2	72	1, 2, 3, 4, 6, 8, 9, 12…	2 × 2 × 2 × 3 × 3
33	1, **3, 11**, 33	3 × 11	73	1, 73	Prime
34	1, **2, 17**, 34	2 × 17	74	1, 2, 37, 74	2 × 37
35	1, **5, 7**, 35	5 × 7	75	1, 3, 5, 15, 25, 75	3 × 5 × 5
36	1, **2, 3**, 4, 6, 9, 12, 18, 36	2 × 2 × 3 × 3	76	1, 2, 19, 76	2 × 2 × 19
37	1, **37**	Prime	77	1, 7, 11, 77	7 × 11
38	1, **2, 19**	2 × 19	78	1, 2, 3, 6, 13, 26, 39, 78	2 × 3 × 13
39	1, **3, 13**, 39	3 × 13	79	1, 79	Prime
40	1, **2**, 4, **5**, 8, 10, 20, 40	2 × 2 × 2 × 5	80	1, 2, 4, 5, 8, 10, 16, 20, 40, 80	2 × 2 × 2 × 2 × 5

Bar models

Decimal place-value grids 1

ten thousands	thousands	hundreds	tens	ones	tenths	hundredths	

ten thousands	thousands	hundreds	tens	ones	tenths	hundredths	

ten thousands	thousands	hundreds	tens	ones	tenths	hundredths	

ten thousands	thousands	hundreds	tens	ones	tenths	hundredths	

Decimal place-value grids 2

hundredths			
tenths			
ones			
tens			
hundreds			
thousands			
ten thousands			

Multiplication and division word problems

1 How many different ways could a class of 30 children be split into groups of equal size?

2 11 drivers are transporting children to a 5-a-side football tournament. Each car can take 4 children. How many 5-a-side teams is this? How many reserve players?

3 For a netball tournament, schools can enter from 30 to 50 players. There are 7 players in each team. How many different numbers of players could they take if they enter complete teams?

4 The population of a town in 1910 was 2953. By 2010 its population had increased 6 times. What was its population in 2010?

5 A farmer has 342 eggs to pack into boxes of six. How many boxes are needed to pack all the eggs?

6 Max has drawn a square on 1 cm squared paper. It is 6 cm high and 6 cm wide. How many 1 cm squares are there inside his diagram?

7 Ceris has built a shape out of cubes. It is 4 cubes high, 4 cubes wide and 4 cubes long. How many cubes has she used?

8 The cost of a school ski trip is £452 per person. What is the total cost for 24 children?

A Long multiplication	B Short multiplication	C Division	D Factors	E Multiples	F Square numbers	G Cube numbers	H Mixed operations

Fractions of 24

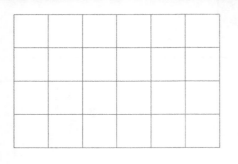
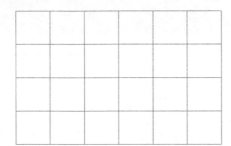
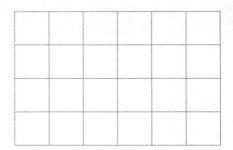

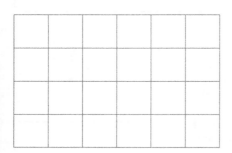
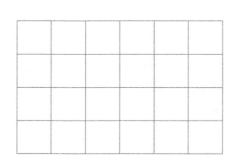
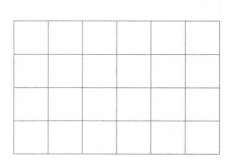

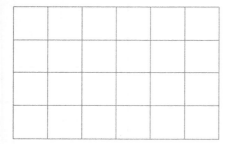
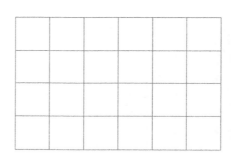
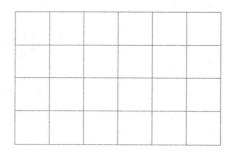

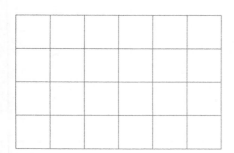
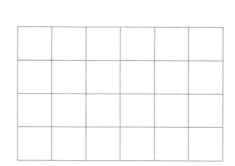
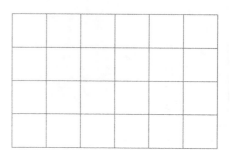

Hundredths

Fraction wall

1									

$\dfrac{1}{2}$ $\dfrac{1}{2}$

$\dfrac{1}{3}$ $\dfrac{1}{3}$ $\dfrac{1}{3}$

$\dfrac{1}{4}$ $\dfrac{1}{4}$ $\dfrac{1}{4}$ $\dfrac{1}{4}$

$\dfrac{1}{5}$ $\dfrac{1}{5}$ $\dfrac{1}{5}$ $\dfrac{1}{5}$ $\dfrac{1}{5}$

$\dfrac{1}{6}$ $\dfrac{1}{6}$ $\dfrac{1}{6}$ $\dfrac{1}{6}$ $\dfrac{1}{6}$ $\dfrac{1}{6}$

$\dfrac{1}{8}$ $\dfrac{1}{8}$ $\dfrac{1}{8}$ $\dfrac{1}{8}$ $\dfrac{1}{8}$ $\dfrac{1}{8}$ $\dfrac{1}{8}$ $\dfrac{1}{8}$

$\dfrac{1}{9}$ $\dfrac{1}{9}$ $\dfrac{1}{9}$ $\dfrac{1}{9}$ $\dfrac{1}{9}$ $\dfrac{1}{9}$ $\dfrac{1}{9}$ $\dfrac{1}{9}$ $\dfrac{1}{9}$

$\dfrac{1}{10}$ $\dfrac{1}{10}$ $\dfrac{1}{10}$ $\dfrac{1}{10}$ $\dfrac{1}{10}$ $\dfrac{1}{10}$ $\dfrac{1}{10}$ $\dfrac{1}{10}$ $\dfrac{1}{10}$ $\dfrac{1}{10}$

$\dfrac{1}{12}$ $\dfrac{1}{12}$ $\dfrac{1}{12}$ $\dfrac{1}{12}$ $\dfrac{1}{12}$ $\dfrac{1}{12}$ $\dfrac{1}{12}$ $\dfrac{1}{12}$ $\dfrac{1}{12}$ $\dfrac{1}{12}$ $\dfrac{1}{12}$ $\dfrac{1}{12}$

Fraction circles

halves

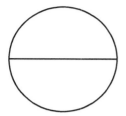 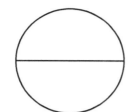 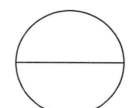 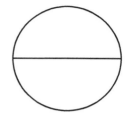

thirds

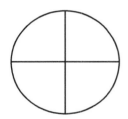 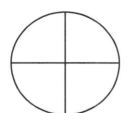 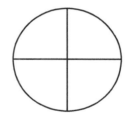 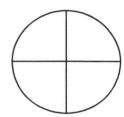

quarters

fifths

Tenths

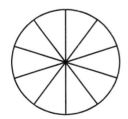 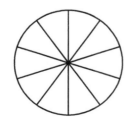 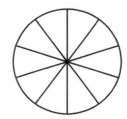 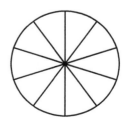

Tenths number lines

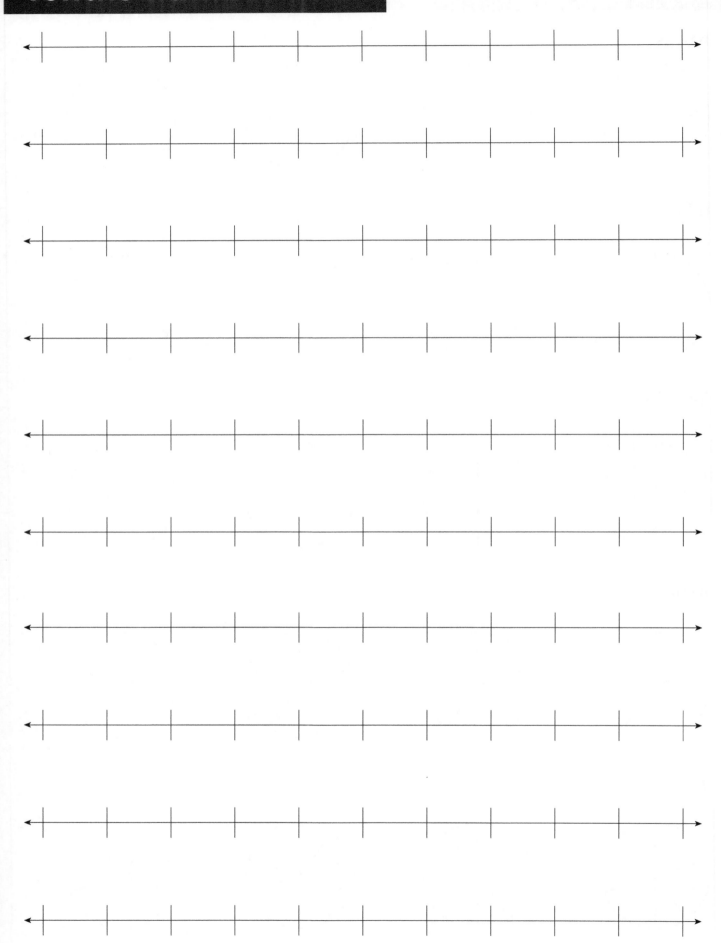

Conversion tables

Imperial measures

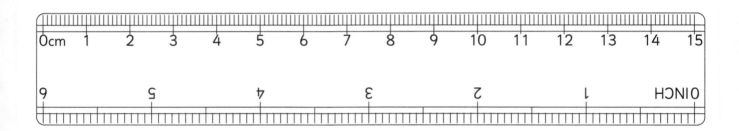

Approximate imperial and metric measure equivalents

Conversions		Imperial measures
Length		**Length**
1 inch = 2.54 cm		12 inches = 1 foot
1 foot = 30.5 cm	1 m = 39 inches	3 feet = 1 yard
1 yard = 0.9 m		1760 yards = 1 mile
1 mile = 1.6 km	1 km = $\frac{5}{8}$ mile	
5 miles = 8 km		
Mass		**Mass**
1 ounce = 28 g		16 ounces = 1 pound
1 pound = 0.45 kg	1 kg = 2.2 pounds	14 pounds = 1 stone
Capacity		**Capacity**
1 fluid ounce = 30 ml		20 fluid ounces = 1 pint
		8 pints = 1 gallon
1 pint = 0.56 litres	1 litre = $1\frac{3}{4}$ pints	
1 gallon = 4.5 litres		

Perimeter and area

Cuboids

a)

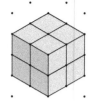

b)

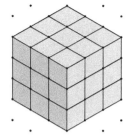

c)

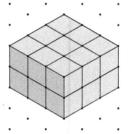

d)

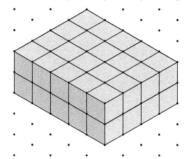

e)

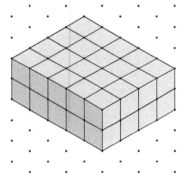

f)

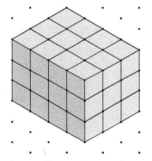

g)

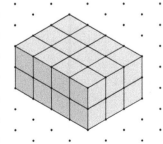

h)

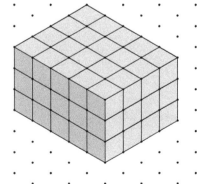

Analogue and digital clock faces

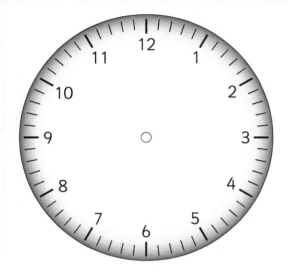

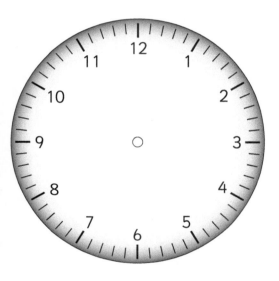

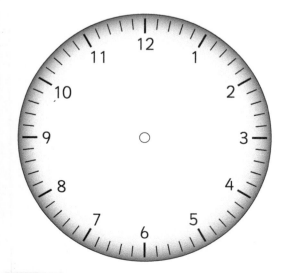

Nets of prisms

triangular prism

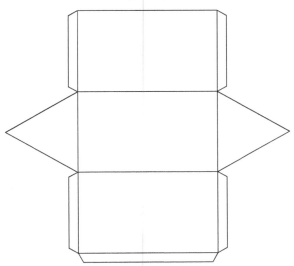

pentagonal prism

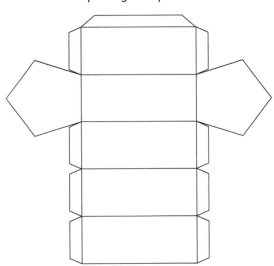

cylinder

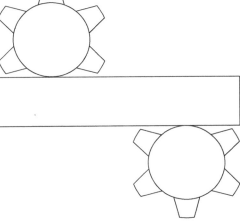

cuboid

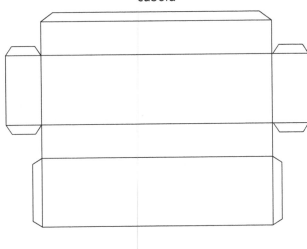

cube

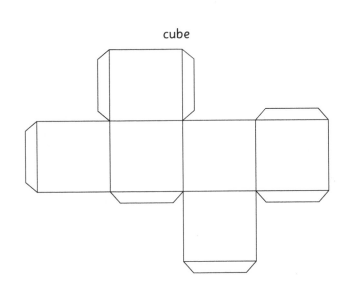

Nets of pyramids and a cone

triangular-based pyramid

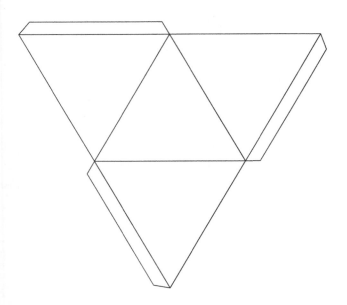

square-based pyramid

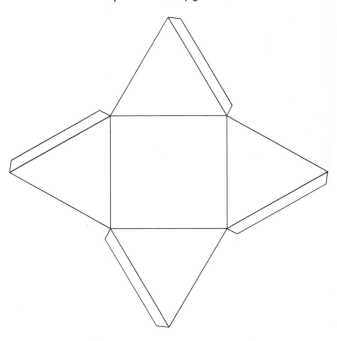

cone

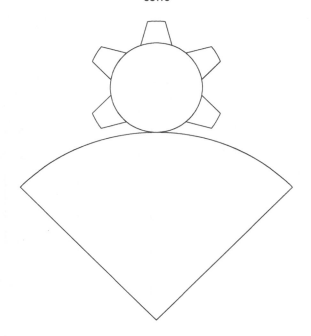

pentagonal-based pyramid

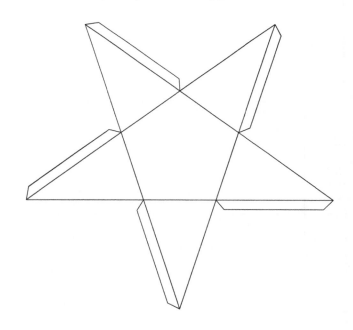

3-D shapes and their nets

triangular-based pyramid	
cone	
cube	
cylinder	
pentagonal-based pyramid	
pentagonal prism	
triangular prism	
cuboid	
square-based pyramid	

2-D shapes

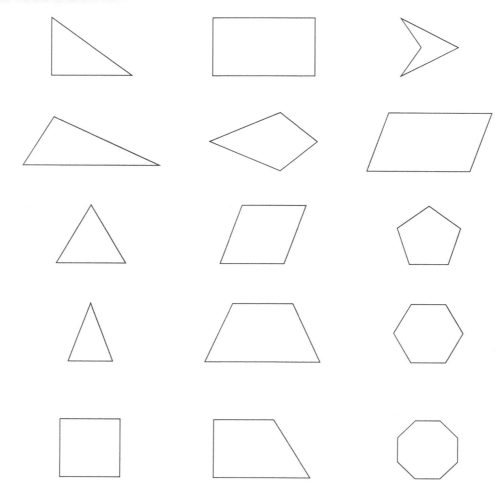

right-angled triangle	scalene triangle	equilateral triangle	isosceles triangle	square
rectangle	kite	rhombus	isosceles trapezium	right-angled trapezium
arrow head	parallelogram	regular pentagon	regular hexagon	regular octagon

2-D shape properties

It has 3 unequal sides. It has 1 right angle.	It has 3 sides. It has 2 equal sides. It has 2 equal angles.	It has 3 non-equal sides. It has 3 non-equal angles.
All angles are (90°). All sides are of equal length. Opposite sides are parallel. The diagonals bisect each other at 90°. The diagonals are equal in length.	Diagonally opposite angles are equal. Opposite sides are of equal in length. Opposite sides are parallel. The diagonals bisect each other.	All angles are equal (90°). Opposite sides are of equal length. The diagonals are equal in length. Diagonals bisect each other. Opposite sides are parallel.
It has 3 equal sides. It has 3 angles of 60°.	Two pairs of adjacent sides are of equal length. One pair of diagonally opposite angles is equal. The diagonals cross at 90°.	Diagonally opposite angles are equal. All sides are of equal lengths. Opposite sides are parallel. The diagonals bisect each other at 90°.
It has 8 equal sides. It has 8 equal angles. It has a regular shape.	It has 1 pair of equal sides. It has 1 pair of equal angles. One pair of opposite sides is parallel.	It has 1 right angle. One pair of opposite sides is parallel.
One pair of opposite sides is parallel. It has 2 pairs of equal sides. It has a reflex angle.	It has 6 equal sides. It has 6 equal angles. It has a regular shape.	It has 5 equal sides. It has 5 equal angles. It has a regular shape.

Coordinate grids

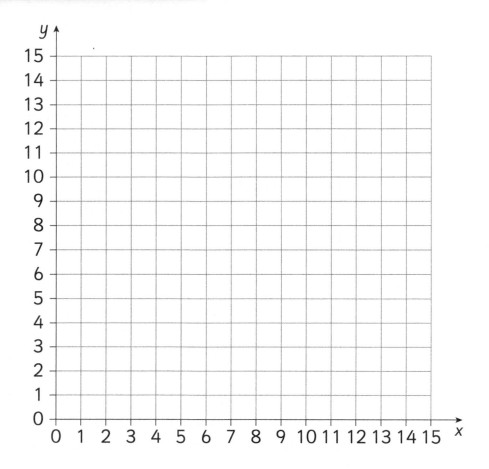

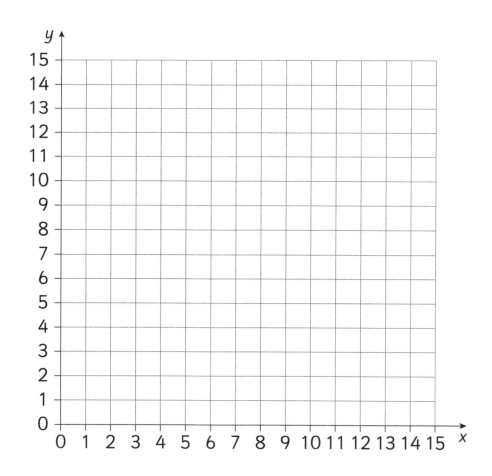

Line graphs

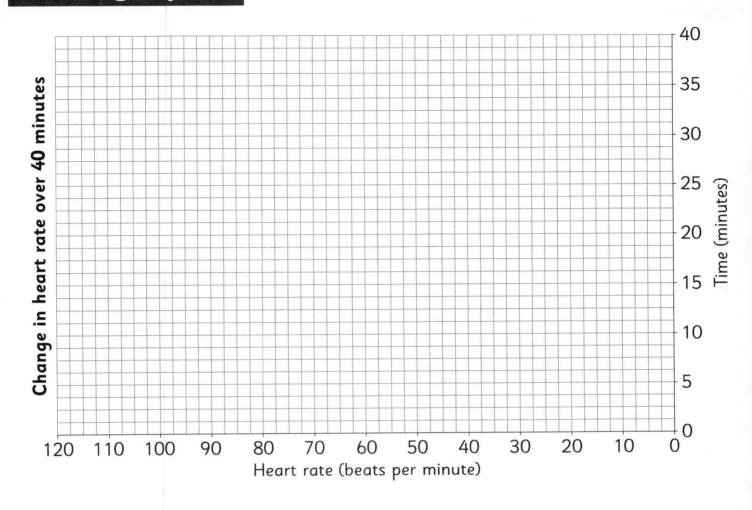

Change in heart rate over 40 minutes

Time (minutes)

Heart rate (beats per minute)

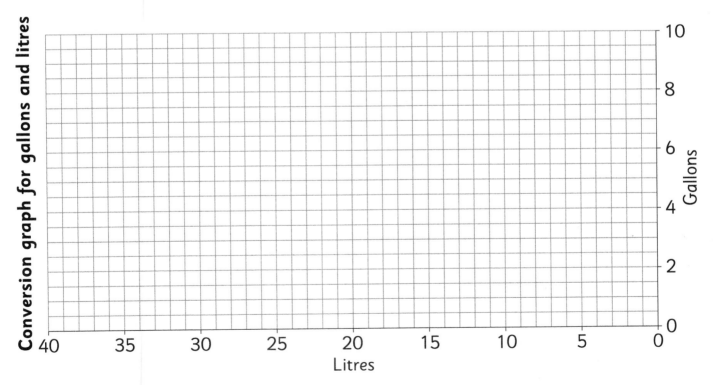

Conversion graph for gallons and litres

Gallons

Litres

Line graph template

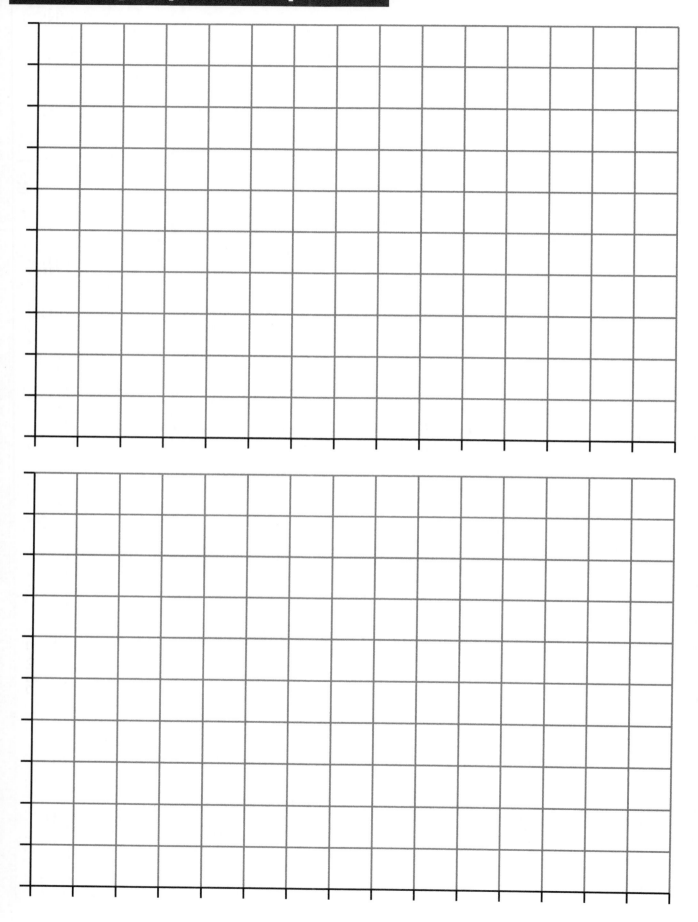

Tables templates

Timetable

Place	time	time	time	time

Timetable

Place	time	time	time	time

Opening times

	Opening times
Monday	
Tuesday	
Wednesday	
Thursday	
Friday	
Saturday	
Sunday	

Two-way table templates

			Total

			Total

				Total

				Total

				Total

Pupil Resource Pack Answers

Unit 1

1. a) 2 475 189   b) 8 763 294   c) 3 589 476
2. a) 21 856   b) 317 401
3. a) one hundred and sixteen thousand, seven hundred and eighty two
   b) seven million, four hundred and eighty thousand, three hundred and twenty-four
4. a) 500   b) 500 000   c) 50 000
5. a) 6 500 005   6 100 365
   6 098 135   5 999 999
   b) 3 587 632   3 587 362
   3 578 326   3 578 263

Unit 1 Quick test

1. a) 4 572 651   b) 5 839 276
2. a) 3 417 680   b) 7 010 049
3. a) five hundred and sixty thousand, one hundred and eight
   b) four million, fifty thousand, seven hundred and ninety
4. a) 300 000   b) 30 000
5. a) 899 998   998 756
   6 999 890   7 275 450
   b) 745 912   745 921
   754 291   754 921

Unit 2

1. a) 4994   5004   5014
   b) 30 000   30 100   30 200
   c) 800 010   800 000   799 990
   d) 100 000   99 900   99 800
2. a) 24 567   b) 33 985   c) 275 392
   d) 882 507   e) 1087   f) 903

Unit 2 Quick test

1. b) 10 000 less   c) 1000 less
2. a) 87 790   87 890   87 990
   b) 21 875   20 875   19 875
   c) 265 600   275 600   285 600
3. a) 24 870   b) 47 921
   c) 77 200   d) 135 892
4. a) 130 300   b) 195 650

Unit 3

1. a) 15°C   b) −10°C
   c) −6°C   d) 17°C
2. a) 0, −5, −10   b) 2, 0, −2
   c) 10, 0, −10   d) 1, −1, −3
3. a) Moscow   b) 22
   c) 14   d) New York

Unit 3 Quick test

1. −9   −3   −1   0   2   12
2. a) −5, 0, 5   b) 0, −9, −18
   c) −1, −4, −7
3. a) −3   b) 7
4. a) St Petersburg
   b) 11   c) 2   d) 32

Unit 4

1. a) 4000   b) 5000
   c) 50 000   d) 40 000
   e) 700 000   f) 800 000
2.

Town	Manchester	Cardiff	Aberdeen	Gloucester	Bath
Population	510 746	335 145	195 021	136 362	94 782
nearest 10	510 750	335 150	195 020	136 360	94 780
nearest 100	510 700	335 100	195 000	136 400	94 800
nearest 1000	511 000	335 000	195 000	136 000	95 000
nearest 10 000	510 000	340 000	200 000	140 000	90 000
nearest 100 000	500 000	300 000	200 000	100 000	100 000

3.
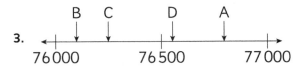

Unit 4 Quick test

1. a) 7000   b) 20 000
   c) 800 000
2. a) 512 950   b) 512 900
   c) 513 000   d) 510 000
   e) 500 000
3. a) 100 000   b) 100 000
   c) 100 000   d) 100 000
4.

Unit 5

1. a) 36   b) 250
   c) 600   d) 2325
   e) 1616   f) 2020
2. a) LXXXVII   b) CXXIII
   c) CCLVIII   d) CDXIX
   e) DCXII   f) DCCCXCV
   g) MCCCLXIV   h) MMMCMXCIX
3. MCMXC   1990
   MMIX   2009
   MCMXCV   1995
   MMX   2010
   MMXIX   2019
   MCMXCIX   1999
4. 8/10/2016
5. Various

Unit 5 Quick test

1. a) 40　　b) 116　　c) 520　　d) 1707
2. a) LXXV　　　　　　b) CCXXXIV
　 c) MD　　　　　　　d) MMDCCLXIII
3. 24/6/1982
4. XXV/XII/MMXV

Unit 6

1. a) Birmingham: one million, eighty-five thousand, eight hundred and ten
   Newcastle: two hundred and sixty-eight thousand and sixty-four
   b) 71 000　　　　　　　　c) 18 093
   d) 1 095 810　　　　　　　e) 69 945
2. Helsinki Bucharest Berlin Glasgow Athens
3. MLXVI 1066 MCCXV 1215
   MDLIX 1559 MMXVIII 2018

Unit 6 Quick test

1. a) eighty-three thousand, one hundred and twenty-five
   b) 83 130　　　　　　c) 80 000
   d) 93 125　　　　　　e) 82 125
2. Answers can be ±50
   A 17 500　　B 17 100　　C 17 750　　D 17 350
3. Depends: 2017 MMXVII 2018 MMXVIII
   2019 MMXIX 2020 MMXX etc
4. C D A E B

Unit 7

1. a) 145 968　　　　　b) 15 026
2. a) 54 778　　　　　 b) 606 684
　 c) 758 616　　　　　d) 141 660
3. a) 14 833　　　　　 b) 45 036
　 c) 5149　　　　　　d) 17 267

Unit 7 Quick test

1. a) 53 172　　　　　 b) 632 439
2. a) 31 017　　　　　 b) 309 709
　 c) 2420　　　　　　d) 44 347
3. a) 18 910　　　　　 b) 15 804

Unit 8

1. a) 630, 460, 710, 680
   b) 480, 160, 290, 70
   c) 5200, 6500, 8300, 9400
   d) 2800, 1500, 4700, −200
2. a) 720/280 480/520 540/460
   　 130/870 910/90 50/950
   b) 3400/6600 1900/8100
   　 9300/700 600/9400
   　 4100/5900 7300/2700
3. a) 544　　　b) 568　　　c) 5363
　 d) 5727　　e) 8724　　f) 3294

Unit 8 Quick test

1. a) 360　　b) 250　　c) 2600　　d) 3900
2. a) 740　　b) 460　　c) 5500
　 d) 800　　e) 3700　　f) 8300
3. a) 627　　b) 277　　c) 6866　　d) 6249
4. a) 490　　b) 530　　c) 3900　　d) 3400

Unit 9

1. < 5000: 1560 + 2945; 2460 + 2399
   > 5000: 2504 + 2504; 3720 + 2940; 8432 − 3109;
   9464 − 3789
2. a) 3000 + 3000 = 6000 too low
   b) 7000 − 3000 = 4000 too high
   c) 15 000 + 13 000 − 14 000 = 14 000 too low
3. 12 050 + 22 945 = 34 995
   34 874 − 6194 = 28 680
   15 896 + 17 914 = 33 810
   61 195 − 29 875 = 31 320
   13 819 + 16 106 = 29 925
4. £300 + £500 + £100 = £900. All rounded up so Oliver is correct.

Unit 9 Quick test

1. £160
2. a) 1000　　b) 8000　　c) 14 000
3. a) incorrect　b) correct　c) incorrect
4. 1000 + 1000 + 200 = 2200. Not enough

Unit 10

1. 66
2. a) various: 9/8 hundreds 6/5 tens, 3/1 units, for example, 963 + 851
   b) 618 − 539 = 79 or 915 − 836 = 79
3. a) A £31 000 B £26 000
   b) £5000
   c) £5725 − £5000 = £725
4. Working out: 17 854 + 49 018 = 66 873,
   70 000 − 66 872 = **3128**

Unit 10 Quick test

1. a) 2534 + 1967　　　　b) 7265 − 4938
2. a) The first class ticket costs £281 more
   b) £5036
3. 960 921 − 613 916 = 347 005

Unit 11

1. a) 3, 6, 9, 12, 15, 18, 21
   b) 4, 8, 12, 16, 20, 24, 28
   c) 9, 18, 27, 36, 45, 54, 63
   d) 11, 22, 33, 44, 55, 66, 77
2. 5, 10, 15, 20, 25, 30, 35, 40, 45, 50, 55, 60
3. a) 40　　　　　　　　b) 70
4. a) 1, 2, 3, 6, 9, 18; 1/18, 2/9, 3/6
   b) 1, 2, 3, 4, 6, 8, 12, 24; 1/24, 2/12, 3/8, 4/6
5. 1, 2, 3, 6
6. a) true　　b) false　　c) true

Unit 11 Quick test

1. 6, 12, 18, 24, 30, 36, 42, 48, 54, 60
2. a) 80　　　　　　　　b) 24
3. a) 1, 2, 5, 10
   b) 1, 2, 3, 5, 6, 10, 15, 30
   c) 1, 2, 5, 10
4. Three of: 1/36; 2/18; 3/12; 4/9; (6/6)
5. a) false　　　　　　b) false
　 c) true　　　　　　d) false
　 e) true

Unit 12

1. **a)** 2, 3, 5, 7, 11   **b)** 4, 10, 12, 15, 18
2. **a)** 7   **b)** 9 or 18
3. $8 = 2 \times 2 \times 2$ $28 = 2 \times 2 \times 7$ $40 = 2 \times 2 \times 2 \times 5$ $30 = 2 \times 3 \times 5$ $15 = 3 \times 5$
4. **a)** 2, 3   **b)** 2, 7   **c)** 2, 3, 5
5. **a)** 2×5 **b)** 3×7 **c)** 3×11
6. It is a multiple of 3, 5; it has 3 and 5 as extra factors

Unit 12 Quick test

1. Prime: 2, 3, 5, 7, 11, 13 Composite: 4, 6, 8, 9, 10, 12, 14, 15 Neither: 1
2. **a)** False   **b)** True
      c) True   **d)** False
3. **a)** 15   **b)** 55   **c)** 12
4. **a)** 2×7 **b)** 5×7 **c)** 2×11
5. Various, for example, it is a multiple of 2, 3, 4, 6 and 12; has factors of 2, 3, 4, 6, 12

Unit 13

1. **a)** 31   **b)** 24   **c)** 47
      d) 66   **e)** 3
2.

	Odd	Even
Prime number	5, 29, 31	2
Composite number	9, 21, 25	18, 22, 28, 30, 32

3. **a)** composite   **b)** multiple
      c) prime factor   **d)** prime
      e) factor

Unit 13 Quick test

1. 2, 3, 5, 7, 11, 13, 17, 19
2. Prime: 23, 31, 37 Composite: 20, 21, 27, 33, 35, 39, 40
3. **a)** It has factors 3 and 11; it is a multiple of 3 and 11
      b) It only has 2 factors, 1 and itself
      c) 2
4. **a)** 17, 19   **b)** 36

Unit 14

1. 76, 38, 19;   52, 26, 13;   36, 18, 9
2. 4, 320, 32 000, 40, 3200, 80, 3200, 4
3. **a)** 2400   **b)** 80
4. **a)** 5   **b)** 9   **c)** 45
      d) 9   **e)** 70   **f)** 40

Unit 14 Quick test

1. **a)** 92   **b)** 78
2. **a)** 42   **b)** 49
3. **a)** 30   **b)** 16000   **c)** 50
4. **a)** 6   **b)** 40   **c)** 11   **d)** 70

Unit 15

1. **a)** 385, 123.4, 18.1, 9.9, 0.45
      b) 7600, 31 700, 45 000, 5670, 187
      c) 894, 28.6, 38.67, 0.72
2. **a)** ÷ 100   **b)** × 10
      c) × 1000   **d)** ÷ 1000

3. 4256   × 10 = 42 560
            × 100 = 425 600
            × 1000 = 4 256 000
            ÷ 10 = 425.6
            ÷ 100 = 42.56
            ÷ 1000 = 4.256

Unit 15 Quick test

1. **a)** 29 600   **b)** 85.7(0)   **c)** 3.958
      d) 634   **e)** 3.2   **f)** 3.76
2. **a)** × 100   **b)** ÷ 1000
      c) × 10   **d)** ÷ 1000
3. 3821

Unit 16

1. **a)** 7638   **b)** 31 182   **c)** 14 015
2. **a)** 76 380   **b)** 311 820   **c)** 140 150
3. **a)** 20 128   **b)** 70 150   **c)** 98 688

Unit 16 Quick test

1. 2112
2. 26 192
3. 32 910
4. 138 667

Unit 17

1. **a)** no   **b)** yes   **c)** yes   **d)** no
2. **a)** 639   **b)** 1849   **c)** 952   **d)** 908
3. **a)** 1458 r 4   **b)** 2065 r 3

Unit 17 Quick test

1. 2369
2. 1367
3. 3068 r 1
4. 784 r 8
5. 8641 is odd; multiples of 4 are even

Unit 18

1. $3 \times 3 \times 3$ — 3^3 — third cube number
   4×4 — 4^2 — fourth square number
   $4 \times 4 \times 4$ — 4^3 — fourth cube number
   $8 \times 8 \times 8$ — 8^3 — eighth cube number
   3×3 — 3^2 — third square number
   8×8 — 8^2 — eighth square number
2. **a)** 5^3 **b)** 7^2
      c) 6^2 **d)** 9^3
3. **a)** 25   **b)** 8
      c) 1000   **d)** 81
      e) 1   **f)** 125
4. **a)** $11^2 = 121$ **b)** $3^3 = 27$
      c) $1^2 = 1$ **d)** $10^3 = 1000$

Unit 18 Quick test

1.

2. a) 11^3 **b)** 12^2
3. a) $6 \times 6 = 36$ **b)** $2 \times 2 \times 2 = 8$
4. a) 49 **b)** 1 **c)** 64 **d)** 64
5. Square: 1, 4, 9, 16, 25, 64, 100
Cube: 1, 8, 27, 64, 1000

Unit 19

1.

Group size	Team Size										
	2	3	4	5	6	7	8	9	10	11	12
20	Y	N	Y	Y	N	N	N	N	Y	N	N
24	Y	Y	Y	N	Y	N	Y	N	N	N	Y
30	Y	Y	N	Y	Y	N	N	N	Y	N	N
36	Y	Y	Y	N	Y	N	N	Y	N	N	Y
40	Y	N	Y	Y	N	N	Y	N	Y	N	N

2. £11 875
3. £286

Unit 19 Quick test

1. a) 125 **b)** 5^3
2. 8 teams and 3 reserves
3. £76
4. 5184

Unit 20

1. a) ÷ / × **b)** − / ×
c) × / − **d)** × / − / +
2. $35 \div 7 = (24 \div 8) + 2$;
$(40 \div 5) + 2 = (5 \times 5) - 15$;
$99 \div 11 = 13 - (12 \div 3)$
3. a) £566: $16 \times 25 = 400$; $16 \times 24 = 384$;
$400 + 384 = 784$; $784 - 218 = 566$
b) £235: C: $152 \times 14 = 2128$; D: $139 \times 17 = 2363$;
$2363 - 2128 = 235$

Unit 20 Quick test

1. a) 16 **b)** 5
2. a) ÷ / − **b)** ÷ / ×
3. Numbers which total 44
4. a) $2 \times 429344 = 858688$
b) 90721: $858688 - 767967 = 90721$

Unit 21

1. a) ÷ 3 × 2 **b)** ÷ 3 **c)** ÷ 4 × 3
d) ÷ 5 **e)** ÷ 5 x 4
2. a) 9 **b)** 7 **c)** 8
d) 12 **e)** 15 **f)** 10
3. a) £9 / £27 **b)** £5 / £15 **c)** £7 / £21
4. 70

Unit 21 Quick test

1. a) $\frac{1}{8}$ **b)** $\frac{3}{5}$
2. a) 6 **b)** 18
3. 20p
4. a) £80 **b)** £24
5. $\frac{2}{3}$ of 90 = 60; $\frac{2}{9}$ of 180 = 40; so $\frac{2}{3}$ of 90 is more

Unit 22

1. a) 2 columns or any 8 squares shaded
b) 1 row or any 6 squares shaded
c) 5 rows or 5 columns or any 50 squares shaded
d) 3 rows or 3 columns or any 30 squares shaded
2. a) $\frac{1}{4}$ or $\frac{4}{16}$ **b)** $\frac{2}{3}$ or $\frac{4}{6}$ or $\frac{24}{36}$
3. a) $\frac{1}{3} \equiv \frac{2}{6} \equiv \frac{4}{12} \equiv \frac{8}{24} \equiv \frac{16}{48}$
b) $\frac{3}{5} \equiv \frac{6}{10} \equiv \frac{12}{20} \equiv \frac{24}{40} \equiv \frac{48}{80}$

Unit 22 Quick test

1. $\frac{2}{3} \equiv \frac{4}{6}$; $\frac{1}{2} \equiv \frac{5}{10}$; $\frac{1}{4} \equiv \frac{2}{8}$
2. a) e.g. $\frac{2}{3} \equiv \frac{4}{6} \equiv \frac{4}{12}$
b) e.g. $\frac{4}{5} \equiv \frac{8}{10} \equiv \frac{16}{20}$
3. a) $\frac{9}{10}$ **b)** $\frac{3}{5}$

Unit 23

1. box 1 $< \frac{1}{2}$: $\frac{1}{3}$ $\frac{4}{9}$ $\frac{3}{8}$ $\frac{5}{12}$ $\frac{3}{10}$
box 2 $= \frac{1}{2}$: $\frac{3}{6}$ $\frac{5}{10}$ $\frac{4}{8}$
box 3 $> \frac{1}{2}$: $\frac{4}{5}$ $\frac{7}{10}$ $\frac{3}{4}$ $\frac{2}{3}$
2. a) $\frac{4}{5} > \frac{7}{10}$ **b)** $\frac{7}{9} > \frac{2}{3}$
c) $\frac{1}{3} < \frac{5}{12}$ **d)** $\frac{1}{4} = \frac{2}{8}$
3. a) $\frac{3}{8}$ $\frac{1}{2}$ $\frac{3}{5}$ **b)** $\frac{2}{5}$ $\frac{1}{2}$ $\frac{5}{6}$
4. a) $\frac{2}{3}$ $\frac{3}{4}$ $\frac{11}{12}$ **b)** $\frac{5}{10}$ $\frac{3}{5}$ $\frac{7}{10}$ **c)** $\frac{1}{6}$ $\frac{1}{3}$ $\frac{4}{9}$
5. Oliver

Unit 23 Quick test

1. a) > **b)** = **c)** <
2. a) $\frac{5}{9}$ **b)** $\frac{1}{4}$ **c)** $\frac{9}{12}$
3. a) $\frac{4}{5}$ $\frac{1}{2}$ $\frac{3}{7}$ **b)** $\frac{1}{2}$ $\frac{3}{8}$ $\frac{1}{4}$ **c)** $\frac{5}{6}$ $\frac{2}{3}$ $\frac{7}{12}$

Unit 24

1. a) 2 circles fully shaded, 1 circle with 4 segments shaded
b) $\frac{14}{5}$
2. a) $\frac{33}{10}$ **b)** $\frac{11}{4}$ **c)** $\frac{7}{3}$
3. a) $4\frac{4}{5}$ **b)** $3\frac{9}{10}$ **c)** $3\frac{1}{2}$
4. a) $\frac{11}{9}$ $1\frac{2}{9}$ **b)** $\frac{7}{5}$ $1\frac{2}{5}$ **c)** $\frac{16}{12}$ $1\frac{4}{12} = 1\frac{1}{3}$

Unit 24 Quick test

1. a) Any whole number and fraction combination

    e.g. $6\frac{2}{3}$ or $1\frac{2}{3}$

    b) e.g. $\frac{12}{5}$ $\frac{6}{8}$

2. $\frac{19}{8}$ $2\frac{3}{8}$

3. $\frac{17}{3}$

4. $6\frac{3}{5}$

5. $\frac{14}{11}$ $1\frac{3}{11}$

Unit 25

1. $=$ $<$ $>$ $=$ $>$ $<$

2. a) $\frac{3}{10}+\frac{8}{10}=\frac{11}{10}$ or $1\frac{1}{10}$

    $\frac{8}{10}-\frac{3}{10}=\frac{5}{10}$ or $\frac{1}{2}$

    b) $\frac{3}{4}+\frac{2}{4}=\frac{5}{4}$ or $1\frac{1}{4}$

    $\frac{3}{4}-\frac{2}{4}=\frac{1}{4}$

    c) $\frac{7}{8}+\frac{2}{8}=\frac{9}{8}$ or $1\frac{1}{8}$

    $\frac{7}{8}-\frac{2}{8}=\frac{5}{8}$

    d) $\frac{8}{12}+\frac{7}{12}=\frac{15}{12}$ or $1\frac{3}{12}$ $1\frac{1}{4}$

    $\frac{8}{12}-\frac{7}{12}=\frac{1}{12}$

3. $\frac{3}{4}-\frac{2}{3}=\frac{9}{12}-\frac{8}{12}=\frac{1}{12}$

4. $\frac{4}{10}+\frac{3}{10}=\frac{7}{10}$

Unit 25 Quick test

1. a) $=1$ b) >1 c) <1

2. a) $\frac{7}{10}$ b) $\frac{5}{8}$ c) $\frac{1}{6}$

3. $1\frac{1}{12}$

4. $\frac{1}{12}$

Unit 26

1. a) $\frac{3}{5}$ b) $3\frac{2}{3}$ c) $\frac{5}{8}$

2. A, F   B, D   C, E

3. a) $\frac{10}{3}$ $3\frac{1}{3}$ b) $\frac{32}{5}$ $6\frac{2}{5}$

    c) $\frac{10}{7}$ $1\frac{3}{7}$ d) $\frac{28}{10}$ $2\frac{8}{10}$

4. a) $9\frac{9}{2}$ or $13\frac{1}{2}$ b) $12\frac{3}{4}$

    c) $8\frac{6}{5}$ or $9\frac{1}{5}$

5. $5\times3\frac{1}{4}=15\frac{5}{4}$ or $16\frac{1}{4}$

Unit 26 Quick test

1. a) $\frac{5}{9}$ b) $\frac{7}{10}$

2. a) $\frac{4}{5}$ b) $\frac{9}{11}$

3. a) $5\frac{1}{4}$ b) $7\frac{1}{5}$ c) $1\frac{4}{8}$ $1\frac{1}{2}$

4. $16\frac{1}{2}$

5. $23\frac{1}{3}$

Unit 27

1. a) $0.1 \ldots 0.9$

    b) $\frac{2}{10}$ $\frac{3}{10}$ $\frac{4}{10}$ $\ldots\frac{9}{10}$ or (using lowest common

    denominators)

    $\frac{1}{5}$ $\frac{3}{10}$ $\frac{2}{5}$ $\frac{3}{5}$ $\frac{7}{10}$ $\frac{4}{5}$ $\frac{9}{10}$

2. a) 0.1 & $\frac{1}{10}$ 0.71 & $\frac{71}{100}$ 0.5 & $\frac{1}{2}$ 0.25 & $\frac{1}{4}$

    0.09 & $\frac{9}{100}$ 0.2 & $\frac{1}{5}$ 0.99 & $\frac{99}{100}$

    b) 0.05 & $\frac{5}{100}$ or $\frac{1}{20}$ 0.4 & $\frac{4}{10}$ or $\frac{2}{5}$

3. a) $\frac{7}{10}$ b) $\frac{83}{100}$ c) $\frac{6}{10}=\frac{3}{5}$

    d) 0.3 e) 0.2 f) 0.01

4. $0.9=\frac{90}{100}$, $90>89$, 0.9 is greater

Unit 27 Quick test

1. a) $0.1 / \frac{1}{10}$ b) $0.01 / \frac{1}{100}$

    c) $0.25 / \frac{1}{4}$ or $\frac{25}{100}$

2. a) $\frac{3}{10}$ b) $\frac{8}{10}=\frac{4}{5}$ c) $\frac{23}{100}$ d) $\frac{7}{100}$

3. a) 0.9 b) 0.6 c) 0.17 d) 0.03

4. $\frac{7}{10}$ is 0.70 as a decimal. 0.65 is smaller

Unit 28

1. a) 0.7; $\frac{7}{10}$; seven tenths

    b) 0.09; $\frac{9}{100}$; nine hundredths

    c) 0.003; $\frac{3}{1000}$; three thousandths

2. a) 0.3 b) 0.37 c) 0.379 d) 0.907

    e) 0.042 f) 0.029 g) 0.001 h) 0.009

3. a) 0.98   **b)** 0.36
4. a) 0.9   **b)** 0.5
5. a) $\dfrac{386}{1000}$ **b)** $\dfrac{215}{1000}$
    c) $\dfrac{640}{1000}$ **d)** $\dfrac{100}{1000}$
6. 0.51 $\dfrac{43}{100}$ 0.429 $\dfrac{428}{1000}$

Unit 28 Quick test

1. a) $\dfrac{3}{100}$, three hundredths

   b) $\dfrac{3}{10}$, three tenths

   c) $\dfrac{3}{1000}$, three thousandths

2. a) 0.007   **b)** 0.043   **c)** 0.387
3. a) 0.75   **b)** 0.1
4. a) $\dfrac{385}{1000}$ **b)** $\dfrac{520}{1000}$ **c)** $\dfrac{700}{1000}$

5. 0.2 0.215 $\dfrac{219}{1000}$ $\dfrac{22}{100}$

Unit 29

1. 4: 4.3, 4.1, 4.4, 4.39, 4.08, 4.27;
5: 4.5, 4.9, 4.6, 4.52, 4.83, 4.72
2. a) 5.7   **b)** 4.3   **c)** 6.8   **d)** 3.2
3. a) 4; 3.6
   b) – **d)** Any three of the following: 3.65: 4, 3.7; 5.36:
   5, 5.4; 5.63: 6, 5.6; 6.35: 6, 6.4; 6.53: 7; 6.5
4. a) $5 + 7 + 8 = 20$ **b)** $5 \times 8 = 40$

Unit 29 Quick test

1. a) 8   **b)** 4   **c)** 10
2. a) 8.2   **b)** 3.6   **c)** 10.0
3. a) 2.9   **b)** 5.0
4. a) $10 \times 9 = 90$ **b)** $24 \div 8 = 3$
5. $6 + 10 + 4 = 20$. All rounded up, so yes, total is less
   than $2 \times £10 = £20$

Unit 30

1. a) >   **b)** >   **c)** <
   d) <   **e)** >   **f)** <
2. 7.789, 7.8, 7.813, 7.82
3. a) Nadia, Sara, Lois, Anna, Patsy
   b) Greg, Tom, Mattieu, Ivan, Tudor

Unit 30 Quick test

1. a) 4.506   **b)** 7.546   **c)** 8.211
   d) 2.93   **e)** 3.6   **f)** 8.512
2. a) 14.5, 14.49, 14.39, 14.345
3. 3.4, 3.445, 3.45

Unit 31

1. a) 4.392   **b)** 8.9   **c)** 5.9   **d)** 9.2
2. a) 6.2   **b)** 4.1   **c)** 6.3   **d)** 2.7
3. a) 1.1   **b)** 0.61   **c)** 0.43   **d)** 1.6
4. a) 0.58   **b)** 2.7   **c)** 0.682   **d)** 2.74
5. 6.5

6. a) 10.325   **b)** 11.128
   c) 10.007   **d)** 7.453
7 a) 2.589   **b)** 2.963
   c) 3.385   **d)** 3.881
8. a) 19.915 km   **b)** 1.045 km
   c) 0.085 km

Unit 31 Quick test

1. a) 4.7   **b)** 2.91
2. 0.75
3. a) 7.154   **b)** 7.9   **c)** 3.3
4. a) Jonathan, Simon, Marty, Andy
   b) 0.202 m
5. 7.4
6. 2.156 m
7. From bottom 6.4, 10.7, 18.3
8. £1.77
9. 0.05 m

Unit 32

1. a) 72%

   b) $1\dfrac{1}{2}$ rows or columns any 15 squares shaded

2. a) $\dfrac{37}{100}$ **b)** $\dfrac{70}{100}$ or $\dfrac{7}{10}$

   c) $\dfrac{6}{10}$ or $\dfrac{60}{100}$ **d)** $\dfrac{1}{100}$

3. a) 0.84   **b)** 0.17   **c)** 0.8   **d)** 0.03
4. 4%, 0.04, $\dfrac{4}{100}$; 40% 0.4, $\dfrac{40}{100}$; 44%, 0.44, $\dfrac{44}{100}$

5. $\dfrac{9}{100}$, 81%, $\dfrac{82}{100}$, 0.825

Unit 32 Quick test

1. a) 39%
   b) 2 columns + 8 cells any 28 squares shaded
2. a) $\dfrac{59}{100}$, 0.59   **b)** $\dfrac{1}{10}$, 0.1   **c)** $\dfrac{6}{100}$, 0.06
3. a) 89%   **b)** 20%   **c)** 3%
4. a) 19%   **b)** 5%   **c)** 80%
5. $\dfrac{73}{100}$, 0.729, 72%, 0.7

Unit 33

1. $\dfrac{1}{2}$, 50% $\dfrac{1}{10}$, 10% $\dfrac{1}{4}$, 25% $\dfrac{7}{10}$, 70% $\dfrac{3}{10}$, 30% $\dfrac{44}{50}$,

   88% $\dfrac{78}{100}$, 78% $\dfrac{3}{4}$, 75% $\dfrac{2}{5}$, 40%

2. a)

Subject	Score	%
Maths	41 out of 50	82%
English	19 out of 25	76%
Art	8 out of 10	80%
Science	15 out of 20	75%
DT	3 out of 5	60%

   b) Maths, Art, English, Science, DT
3. 44%

Unit 33 Quick test

1. a) $\frac{1}{2}$ b) $\frac{1}{4}$ c) $\frac{7}{10}$
2. a) 10% b) 60% c) 75%
3. 76%
4. 90%
5. 28%

Unit 34

1. a)

kilograms	grams
1	1000
2	2000
5	**5000**
0.5	500
3.5	**3500**
5.5	5500
0.4	**400**
0.25	250

b)

metres	centimetres
1	100
2	**200**
5	500
2.5	**250**
4.5	450
7.5	750
0.1	10
0.8	**80**

2. a) 300 / 3 or 3000/30
   b) 3000 / 3
   c) 30 / 3 or 300/30 or 3000/300
   d) 3 / 3000
3. a) ÷ 100 b) × 1000
   c) ÷ 1000 d) ÷ 10
4. a) 6 b) 4 c) 3500 d) 7.5
   e) 500 f) 75 g) 20 h) 0.6

Unit 34 Quick test

1.

litres	millilitres
1	1000
2	2000
0.5	**500**
4	**4000**
4.5	4500
0.25	**250**
0.75	750

2. a) 100 b) 1000 c) 10 d) 1000
3. ÷ 1000
4. a) 6 b) 5 c) 2500
   d) 3.5 e) 3.6 f) 0.7

Unit 35

1.

	length	mass	capacity
imperial	mile, inch, foot	pound, ounce	gallon, pint
metric	kilometre, metre, centimetre	gram, kilogram	litre, millilitre

2. a) 16 b) 15 c) 80
   d) 250 e) 200

3. a) 4.5 kg b) 7 pints
4. 448 g
5. 1 m 65 cm

Unit 35 Quick test

1. metric: kilogram, millimetre, gram
   imperial: pint, mile, pound
2. a)

fluid ounce	ml
1	30
2	60
5	150
10	300
20	600
50	1500
100	3000

   b) i) 120 ii) 450 iii) 3 iv) 11
3. The baby boy.
   3.5 kg (3.5 × 2.2) = 7.7 pounds > 7.5 pounds or
   7.5 pounds
   (0.45 × 7.5) = 3.375 kg < 3.5 kg

Unit 36

1. 'The length of the outside of a shape' or similar
2. a) P = 6 + 6 + 6 + 6 = 4 × 6 = 24 cm
   b) P = 6 + 4 + 6 + 4 = 2 × (6 + 4) = 20 cm
3. a) 16 cm b) 18 cm
4. 68 cm

Unit 36 Quick test

1. a) P = 3 + 6 + 3 + 6 = 2 × (3 + 6) = 18 cm
   b) P = 4 + 4 + 4 + 4 = 4 × 4 = 16 cm
2. 20 m
3. 36 m

Unit 37

1. a) 21 cm² b) 49 cm²
2. Farmer Brown: FB 64 m², FH 54 m²
3. 19 to 21 cm²

Unit 37 Quick test

1. 81 cm²
2. rectangle 30 cm²; square 64 cm²
3. 22 to 24 cm²

Unit 38

1. a) = f), b) = d) and c) = e)
2. a) halfway between second and third marks
   b) 750 ml
3. 300 to 500 ml
4. 16 to 20 cm³

Unit 38 Quick test

1. a) 30 cm³ b) 24 cm³
2. halfway between fourth and fifth marks
3. a) 1 litre b) 840 to 860 ml
4. 780 to 800 ml

Unit 39

1. a) 21 b) 29 c) 26
   d) 15 e) 20 f) 30
2. 3 hours 30 minutes
3. 5 days
4. 3:10
5. 3 hours 55 minutes

Unit 39 Quick test

1.
2. a) 4 b) 42 c) 165
3. 3 minutes 15 seconds
4. quarter to 10; 9:45
5. 3 hours 20 minutes

Unit 40

1. 1.45 l = 1 litre 450 ml; 1.405 l = 1405 ml;
   1.4 l = 1400 ml; 0.4 l = $\frac{2}{5}$ litre; 0.45 l = 450 ml
2. 70 cm, 0.72 m, 74 cm, $\frac{3}{4}$ m, 1 m
3. £7
4. £1.20
5. a) 25 b) £5
6. a) 240 g b) 1 c) 40 ml d) 32
7. a) $\frac{1}{4}$ b) $\frac{3}{4}$ c) $\frac{1}{10}$ d) $\frac{3}{5}$
8. a) 500 ml b) 750 g c) 100 m
   d) 20 cm e) 170 cm f) 2250 g
9. 1.85 m
10. £3.80
11. 5.5 litres

Unit 40 Quick test

1. a) 100 ml b) 50 cm
   c) 750 m d) 1900 g
2. a) $\frac{3}{10}$ b) $\frac{1}{4}$
3. 15
4. a)

cookies	sugar		cookies	flour		cookies	milk
10	100 g		10	200 g		10	50 ml
5	50 g		5	100 g		5	25 ml
20	200 g		20	400 g		20	100 ml
1	10 g		1	20 g		1	5 ml
25	250 g		15	300 g		40	200 ml

   b) 300 g c) 250 g d) 40
5. 1.85 kg or 1850 g
6. £1.37
7. 4.5 kg = 3 × 1.5 kg packets
8. 8

Unit 41

1. A, D, E
2. a) triangular based prism

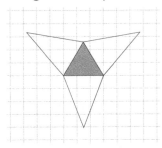

   b) hexagonal based prism

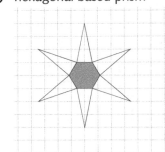

3.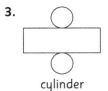

   cylinder   pentagonal-based   pentagonal
               pyramid   prism

Unit 41 Quick test

1. Square based pyramid

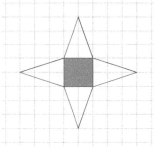

2. cuboid

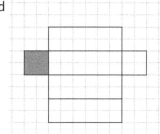

3. a) cube
   b) triangular prism
   c) cone

Unit 42

1. a) obtuse b) reflex c) acute d) right
   e) reflex f) acute g) obtuse h) right
2. a) A right B acute C obtuse D right
   b) P acute Q reflex R acute
3. Various

Unit 42 Quick test

1. **a)** right **b)** 135° **c)** 45° **d)** 200°
2. A acute B obtuse C right D reflex
3. **a)** and **b)** various
   c) on diagonal perpendicular to given line

Unit 43

1. **a)** 60° **b)** 145° **c)** 105° **d)** 38°
2. **a)** 40° **b)** 115° **c)** 145°
   d) 78° all +/− 2 degrees
3. A = B = 105° (+/− 2)
   C = D = 75° (+/− 2)
4. Check accuracy of pupil's diagrams

Unit 43 Quick test

1. **a)** 115° **b)** 42°
2. **a)** 130° **b)** 45° (+/− 2°)
3. Check accuracy of pupil's drawings
4. A = 130°, B and C = 25°

Unit 44

1. **a)** 60° and 120° **b)** 135° and 45°
2. **a)** 40° **b)** 135°
   c) 65°
3. **a)** 95°/ 85°/ 95° **b)** 105°
   c) 70°

Unit 44 Quick test

1. 72 + 98 = 170, not 180
2. A = 55°
3. 35°
4. $p = 54°$, q and $r = 126°$
5. 130°

Unit 45

1.
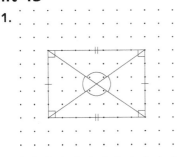

2. **a)** 8 cm and 6.7 cm
   b) a = 80°, b = 100°, c = 80°
3. 15 cm
4. 10 cm

Unit 45 Quick test

1. Opposite sides are parallel; Opposite sides are equal; All angles are 90°
2. **a)** 12 cm and 6.25 cm
   b) d = 125°, e = 55°, f = 125°
3. 20 cm
4. 5 cm

Unit 46

1.

shape	1 or more right angles	opposite angles are equal	more than 4 sides	1 or more pairs of parallel sides
	✓	✗	✗	✓
	✗	✗	✓	✗
	✓	✓	✗	✓
	✗	✓	✓	✓

2.

	regular	irregular
parallel sides	square	rhombus
no parallel sides	equilateral triangle	kite

3. Diagram of parallelogram or rhombus

Unit 46 Quick test

1. Regular: square, equilateral triangle, regular pentagon, regular hexagon
   Irregular: rhombus, irregular hexagon, irregular pentagon, isosceles triangle

2.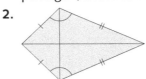

3.

	90° angle(s)	no 90° angles
2 pairs of equal sides	rectangle	**parallelogram**
1 pair of parallel sides	**right angled** trapezium	irregular hexagon

4. Diagram of isosceles triangle

Unit 47

1. **a)**
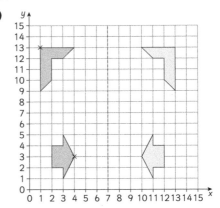

b) (13, 13); (10, 3)

2. a)

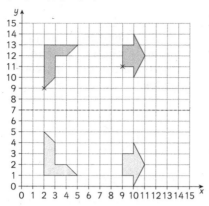

b) (2, 5); (9, 3)

3. a) to **d)**

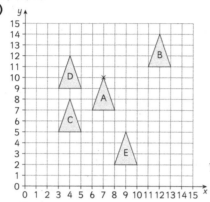

4. a) 5 left 4 down
  b) 3 right 2 up
  c) 8 left 6 down
  d) 5 right 7 down

5. B (12, 14)   C (4, 8)   D (4, 12)   E (9, 5)

Unit 47 Quick test

1.

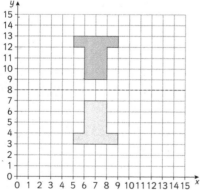

2.

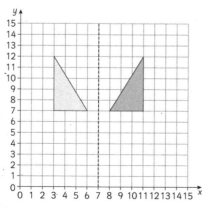

3. a) and **b)**

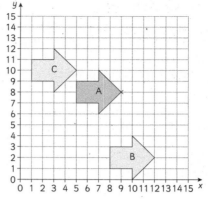

  c) B (12, 2); C (5, 10)
4. a) 4 left, 3 up   **b)** 4 right 5 down

Unit 48

1. a) 2 inches = 5 cm
  b) 15 cm   **c)** 4"   **d)** 7"
2. a) 25 cm   **b)** 12"
3. a) 4" = 10.5 cm so 11 cm longer
  b) 11" = 27.5 cm so 11" shorter
4. a) 50 cm   **b)** 7
  c) 80–90 cm   **d)** 5 and 6
  e) 11, line flattens out, growth stays the same
  f) 1–6 approx 175 cm +; (6–12 approx 75 cm)

Unit 48 Quick test

1. a) 8 km   **b)** 15 miles
  c) 19–20 km   **d)** 17–18 miles
2. a) 80 km   **b)** 30 miles
3. a) 15 miles (= 24 km)
  b) 60 mph = 96 km/h so 100 km/h is faster
4. a) 90°C   **b)** 20°C
  c) 45 minutes   **d)** about 50°C
  e) about 20 minutes   **f)** 40 degrees
  g) 0–15 minutes

Unit 49

1. a) 51 minutes   **b)** 7 minutes
  c) Shurdington/Cheltenham, 18 mins
  d) 09:27   **e)** 10:48

2.

	French	Spanish	German	total
Boys	13	22	27	62
Girls	26	15	17	58
	39	37	44	120

  a) 44   **b)** 35

Unit 49 Quick test

1. a)

	Y4	Y5	Y6	Total
Boys	14	15	17	46
Girls	16	15	13	44
	30	30	30	90

  b) 44
2. a) 56 minutes   **b)** 47 minutes
  c) The 13:44 train takes 1 hour 56 minutes.
    The 14:45 train takes 2 hours 3 minutes.
  d) 17:03